GRAND STRATEGIES

in War and Peace

GRAND STRATEGIES

in
War and Peace

Edited by
PAUL KENNEDY

YALE UNIVERSITY PRESS
New Haven and London

Designed by Sylvia Steiner. Set in Melior type by
The Composing Room of Michigan, Inc.
Printed in the United States of America by Vail-Ballou Press,
Binghamton, New York.

Library of Congress Cataloging-in-Publication Data

Grand strategies in war and peace / edited by Paul Kennedy.
   p.   cm.
   Includes bibliographical references and index.
   ISBN 0-300-04944-7
   1. Strategy—Case studies.   2. Strategy—History.   3. Europe—
History, Military.   4. Europe—National security—History.
5. United States—Military policy.   6. United States—National security.   I. Kennedy, Paul M., 1945- .
U162.G68   1991
355.02'0722—dc20   90-23410
                       CIP

10 9 8 7 6 5 4 3 2 1

Contents

Preface

The chief purpose of the essays which follow is to present the reader with historical case studies of "grand strategy"; that is to say, with assessments of the success or failure with which various powers of Europe sought to integrate their overall political, economic, and military aims and thus to preserve their long-term interests. The chapters in this book focus upon some of the classic examples of European grand strategy, from ancient Rome to Churchill's Britain, from imperial Spain to the Soviet Union today. Unlike many other studies of strategy, these essays are often as concerned with the non-military as with the military aspects of national policy and, in many cases, with the evolution of grand strategy in peacetime as well as in wartime.

The second purpose of this work is more contemporary than historical. It relates to the debate that is currently taking place about the proper balance of priorities—in other words, the grand strategy—that should be carried out by the United States in the world today. Although in many respects different from the traditional European great powers, the U.S. also has to evolve policies that will ensure its long-term interests in a complex and swift-changing world; in consequence, it needs to integrate its political, economic, and military aims in a coherent fashion, for years of peace as well as the possibility of war. The two essays by the editor, one preceding and one following the historical case studies, are an attempt to relate those European experiences to the American position today.

Paul Kennedy

Grand Strategy in War and Peace: Toward a Broader Definition

Paul Kennedy

In much of the older literature on the nature of armed forces and warfare, a simple division was made between two levels of analysis: tactics and strategy. Such a division appeared perfectly straightforward, especially to those engaged in training future battlefield commanders at the military academies. "Tactics," as Clausewitz had proposed, "is the art of using troops in battle; strategy is the art of using battles to win the war."[1] And what could be more important to military men than deploying troops and winning wars?

Like most simple definitions, however, this one also required modification—and further subdivision. For example, tactics itself might be divided into the actual techniques of fighting by the troops (infantry squares, combined-arms), and into the maneuvering of the entire army or the fleet by the commander—"grand tactics," as it is sometimes referred to.[2] According to other authorities, the conduct of a single battle or campaign—Blenheim, say, or Gallipoli—is best described as taking place at the *operational* level, midway between the tactical and the strategic.[3]

Just as tactics can be analyzed and understood at various levels, so also can strategy. One use of the word would be almost purely military in its focus, as in, say, General Sir Douglas Haig's strategy on the western front, or General Douglas MacArthur's "island-hopping" strategy in the Pacific. As such, it leaves little or no room for the consideration of the nonmilitary dimensions of conflict, or for the longer-term and *political* purposes of the belligerent state as a whole. To cover those dimensions, therefore, military writers have suggested that the most appropriate term to employ is *grand strategy*. In Edward Mead Earle's words, "strategy is the art of controlling and utilizing the resources of a nation—or a coalition of nations—including its armed forces, to the end that its vital interests shall be effectively promoted and secured against enemies, actual, potential, or merely presumed. The highest type of strategy—sometimes called grand strategy—is that which so integrates the policies and armaments of the nation that the resort to war is either rendered unnecessary or is undertaken with the maximum chance of victory."[4] By such a definition, Earle massively extended the realm of enquiry about "grand strategy" to encompass national policies in peacetime as well as in wartime. But perhaps even that was not as radical in its implications as the argument advanced by the military writer Sir Basil Liddell Hart in his book *Strategy*, where he proposed that since "the object in war is to obtain a better peace—even if only from your own point of view— . . . it is essential to conduct war with constant regard to the peace you desire." It therefore followed that "if you concentrate exclusively on victory, with no thought for the after-effect, you may be too exhausted to profit by the peace, while it is almost certain that the peace will be a bad one, containing the germs of another war."[5]

This argument led Liddell Hart to draw two broad conclusions. The first was a widening of the meaning of the word *victory*, since (he felt) "victory *in the true sense* implies that the state of peace, and of one's people, is better after the war than before. Victory in this sense is only possible if a quick result can be gained or if a long effort can be economically proportioned to the national resources. The end must be adjusted to the means."[6]

The second conclusion, following directly from Liddell Hart's belief that the key task facing national decision makers was to relate ends and means, was that "grand" strategy had to involve much more than the supervision of battles. On the contrary,

> Grand Strategy should both calculate and develop the
> economic resources and manpower of nations in order to
> sustain the fighting services. Also the moral resources—
> for to foster the peoples' willing spirit is often as import-
> ant as to possess the more concrete forms of power.
> Grand Strategy, too, should regulate the distribution of
> power between the several services, and between the

> services and industry. Moreover, *fighting power is but*
> *one of the instruments of grand strategy*—which
> should take account of and apply the power of financial
> pressure, of diplomatic pressure, of commercial pres-
> sure, and, not least of ethical pressure, to weaken the
> opponent's will. . . . It should not only combine the
> various instruments, but so regulate their use as to avoid
> damage to the future state of peace—for its security and
> prosperity.[7]

Although those words are rather abstract and general, they were not formulated in a vacuum. Liddell Hart's definition of grand strategy was, as many later observers pointed out, intimately connected with his own personal circumstances.[8] After being gassed and wounded on the Somme in 1916, he later developed a strong criticism not only of Haig's frontal offensives but also of Britain's World War I strategy of the "continental commitment." To restore battlefield mobility, he urged the idea of the primacy of the "indirect approach," thereby becoming forever associated with the intellectual origins of blitzkrieg warfare. Moreover, Liddell Hart's dislike of Britain's overcommitment (as he viewed it) to the western front led him to put the case for "the British way in warfare": that is to say, for the "historical strategy" of an island-state chiefly reliant upon sea power, and contributing the in-struments of the maritime blockade, financial subsidies, and pe-ripheral operations—but *not* a large-scale continental army—to the coalition assembled to defeat any power which sought to dominate Europe by force. Such a grand strategy was economical (in terms of both Britain's manpower and resources), it was calculated and moder-ate, and it involved a constant assessment of means and ends, as had occurred in the eighteenth-century struggles against France. The Brit-ish strategy of 1914–1918 flouted that tradition and, because it cost too much, meant that the nation and its people were not better off in "victory" than they had been previously.[9]

Over the past two decades, Liddell Hart's strategical diagnosis has come in for considerable criticism. It displayed, various writers aver, a nostalgia for an eighteenth-century mode of warfare which had be-come much less efficacious in the twentieth century, when neither maritime blockades nor peripheral operations could bring down Wilhelmine (let alone Nazi) Germany; only a full-scale, unrelenting, and costly "continental commitment" could ensure that. And if Lid-dell Hart's proposals were somewhat defective in respect of British strategy, they were far less appropriate—and useful—for *other* na-tions (for example, Poland) to adopt.[10]

But if Liddell Hart's ideas about British strategy remain debatable, his contribution to the study and understanding of grand strategy as a whole was very important. What he and, slightly later, Earle were arguing for was a substantial broadening of the definition of the term, to show what a complex and multilayered thing a proper grand strategy had to be—and thus to distinguish it very firmly from the strictly operational strategy of winning a particular battle or campaign.

Just how broad that definition has become is worth some further reflection. To begin with, a true grand strategy was now concerned with peace as much as (perhaps even more than) with war. It was about the evolution and integration of policies that should operate for decades, or even for centuries.[11] It did not cease at a war's end, nor commence at its beginning. This was, Liddell Hart observed, the real point of Clausewitz's observation that war was "a continuation of policy by other means."[12]

Second, grand strategy was about the balancing of ends and means, both in peacetime and in wartime. It was not enough for statesmen to consider how to win a war, but what the *costs* (in the largest sense of the word) would be; not enough to order the dispatch of fleets and armies in this or that direction, but to ensure also that they were adequately provided for, and sustained by a flourishing economic base; and not enough, in peacetime, to order a range of weapons systems without careful examination of the impacts of defense spending. It is true that Liddell Hart himself showed little interest in the financing of war, or even in such a critically important field as the logistics of war;[13] but in his emphasis upon whether a war paid or whether victory could have been achieved at a lesser cost, and especially in his highlighting of the economic purposes and underpinnings of Britain's traditional policy, he pointed to components of grand strategy which later historians have come to regard as central.[14]

Third, because this broader definition comprehends much more than what happens on the battlefield itself (more, even, than what is happening amid the armed forces themselves), the student of grand strategy needs to take into consideration a whole number of factors that are not usually covered in traditional military histories, including:

1. The critical importance of husbanding and managing national resources, in order to achieve that balance between ends and means touched upon above. As the recent historiography of the early modern European state now makes clear, the juggling of scarce resources was the constant preoccupation of monarchs and statesmen, and the single most important factor in explaining defeat or victory.[15]

In the era of industrial and technological warfare, the economic component to grand strategy occupies a no less critical place.

2. The vital role of diplomacy, in both peacetime and wartime, in improving the nation's position—and prospects of victory—through gaining allies, winning the support of neutrals, and reducing the number of one's enemies (or potential enemies). It is, for example, difficult to imagine the brilliantly swift unification of Germany under Prussian leadership in the 1860s without Bismarck's success in diplomatically isolating Berlin's successive foes. It was also through the weapon of diplomacy (tightening the alliance with Russia, winning over Italy and Spain, forging the Entente Cordiale with Britain) that the French Foreign Ministry under Theophile Delcassé helped to "compensate" for its economic and military inferiority to Germany after 1900.[16] Conversely, a clumsy diplomacy like that of Germany under Wilhelm II or Brezhnev's Soviet Union can all too frequently weaken a country's grand-strategic position.

3. The issue of national morale and political culture, which is of importance not only on the battlefield but also in a population's willingness to support the purposes and the burdens of the war—or the cost of large defense forces in peacetime. The lack of enthusiasm among much of the Italian population for Mussolini's military ventures, as compared with the Japanese conviction that death was the only honorable alternative to victory, provide good modern examples of the importance of these nonmaterial factors. But so also, more recently, does the story of United States involvement in the Vietnam War, where the massive commitment of men and resources was not sustained by the popular support of the American nation—in contradiction of Clausewitz's belief that the *political* component of war was central, and in disregard of Liddell Hart's observation that fostering "the people's willing spirit is often as important as [possessing] the more concrete forms of power."[17]

The crux of grand strategy lies therefore in *policy*, that is, in the capacity of the nation's leaders to bring together all of the elements, both military and nonmilitary, for the preservation and enhancement of the nation's long-term (that is, in wartime *and* peacetime) best interests. Such an endeavor is full of imponderables and unforeseen "frictions." It is not a mathematical science in the Jominian tradition, but an art in the Clausewitzian sense—and a difficult art at that, since it operates at various levels, political, strategic, operational, tactical, all interacting with each other to advance (or retard) the primary aim.

The history, geography, and culture of each country on our planet are unique—just as each war is different, and each battle particular unto itself—but there are always some unifying elements, deriving from our common humanity. One of them is the demand placed upon the *polities* of this world, whether ancient empires or modern democracies, to devise ways of enabling them to survive and flourish in an anarchic and often threatening international order that oscillates between peace and war, and is always changing. Given all the independent variables that come into play, grand strategy can never be exact or fore-ordained. It relies, rather, upon the constant and intelligent reassessment of the polity's ends and means; it relies upon wisdom and judgment, those two intangibles which Clausewitz and Liddell Hart—despite their many differences—esteemed the most. Finally, we need to understand that wisdom and judgment are not created in isolation; they are formed, and refined, by experience—including the study of historical experiences.

It is with this in mind that the following case studies of grand strategy are presented. They are all located in the European historical experience and they all, rather naturally,[18] concern "great" rather than "small" or medium-sized powers. Because Liddell Hart's own concepts about grand strategy derived from his study of the British experience from the sixteenth to the twentieth centuries, three essays focus upon that nation. The authors—John Hattendorf, Michael Howard, and Eliot Cohen—use the context of a great wartime struggle to illustrate the British efforts to evolve a grand strategy which would integrate all the necessary strands of policy, the European and the extra-European, the military and the diplomatic, the economic and the political. Despite all the changes over time, they confirm an essential continuity, of both problems and "solutions," in the story of British grand strategy. Whether in Marlborough's time or in Churchill's, decision makers in London faced much the same scene: alliances had to be preserved, but the often unwelcome and distorting consequences of those ties needed to be minimized; the pursuit of all-out victory had to be set against the costs of a premature or over-ambitious campaign; tactical incompetence and operational setbacks constantly threatened the cleverest strategical stroke; the lack of resources, or the constraints of being heavily committed to theaters which it was impossible to abandon, compelled Britain to conduct war (in Lord Kitchener's words) "as we must, not as we should like"[19]—and often made a nonsense of having fixed strategic blueprints. Again and again, therefore, one is struck by the importance of flexibility and frequent self-assessment, by the need for broad vision and balance, and by the centrality of the political dimension in grand strategy.

The three essays on British wartime grand strategy are followed, and complemented, by five further essays (by Arther Ferrill, J. H. Elliott, Dennis Showalter, Douglas Porch, and Condoleezza Rice), each of which focuses upon an individual great power. The countries themselves were deliberately chosen to provide the reader with a set of case studies widely different both in time and place: the Roman Empire, with perhaps the most successfully (if subconsciously) sustained "grand" strategy in history; imperial Spain in the early modern era; Germany and France—so close together, yet so different in their strategies—in the nineteenth and twentieth centuries; and the Soviet Union, yesterday and today. Once again, in all cases the emphasis is not simply upon winning battles, or even upon winning wars, but upon the broad political[20] circumstances in which victory should be set. In all cases, the intention is to follow Clausewitz and Liddell Hart in conceiving of grand strategy in the widest possible terms.

The collection concludes with the editor's reflections upon American grand strategy, today and tomorrow. This seems appropriate for a number of reasons. The United States today in many respects occupies a position similar to that of earlier "number one" great powers like Rome, Spain, and Britain.[21] It faces challenges to its interests in different parts of the globe; there are never enough armed forces to be safe everywhere, and priorities have to be set; even in peacetime, it has to think about and prepare for war—although not to the extent that such preparations would weaken its long-term power; and it needs to integrate the military dimensions of its grand strategy with nonmilitary diplomacy, technology, and culture. All this must take place in a world of constant flux, where it is exceedingly difficult to distinguish ephemeral happenings from those which will alter the strategic landscape.

The purpose of this collection is therefore twofold. First, to present a set of historical case studies of grand strategy, which synthesize and survey the experiences of the most important of the European great powers, from classical times to the present. As such, they can be used in the course of academic study, or for general erudition.

The second purpose is to use these case studies, as Churchill himself used history, for instruction.[22] No doubt the circumstances in which American policymakers find themselves at the close of the twentieth century are special unto themselves. There is also no doubt that history does *not* repeat itself, in the very narrow sense of the term. But even as America's leaders seek to evolve their own "integrated, long-term strategy,"[23] they ought to do so with an awareness of history, and with an understanding of those features of grand strategy which exist at all times, and in all countries.

The British Way
in Warfare

2

Alliance, Encirclement, and Attrition: British Grand Strategy in the War of the Spanish Succession, 1702–1713

John B. Hattendorf

In thinking about British strategy in the War of the Spanish Succession, the modern student is faced with several problems. First is the long-standing problem that the events of the war have been described largely in terms of personalities. This is particularly true of studies in English which are dominated by biographies of the duke of Marlborough and by the use of Marlborough's published personal correspondence as the principal source on England's conduct of the war. Second, the bulk of historians working on this period have been interested in domestic party politics rather than in questions of foreign policy. Although there are some important exceptions to these trends,[1] they have resulted in the standard textbook view that ministers of state did not think seriously and coherently about grand strategy in the eighteenth century, but rather left it to men like Marlborough.

This view has persisted partly because the student of grand strategy will find no single collection of documents which preserves evidence of the assumptions, ideas, and purposes relating to England's contribution in the War of the Spanish Succession. One longs for a detailed series of full cabinet minutes or the discovery of the secret papers of a strategic directorate in Whitehall. One is left, however, with only the barely legible scribblings of an occasional minister jotting notes for himself at a cabinet meeting, a mountain of orders and instructions, legions of reports, the humdrum routine of interdepartmental correspondence, and the bits and pieces of personal letters strewn in record offices and libraries across Britain, in America, and in diplomatic archives on the Continent.

If one is to know anything of England's grand strategy in this war, it seems that it must come through the process of deduction while sifting through the papers of those who actively participated in the process by which grand strategy was made and carried out. If Marlborough's comprehension, as Winston Churchill claimed, "extended to all theatres and his authority alone secured design and concerted action . . . [h]e was . . . though a subject virtually master of England."[2] If Churchill is correct, then the standard view is probably accurate, but the documents suggest something different.

In the first place, such a characterization suggests something which is quite different from the larger pattern of British constitutional development and would seem to be an exception to the rule rather than part of that evolving process. There is no doubt that Marlborough was an important man and made great contributions, but a study of the documents shows that he worked within the framework of a complex governmental machine that allowed no single element within it to reign supreme. As Sir William Blackstone characterized the relationship between various parts of English government, they are like "distinct powers in mechanics, they jointly impel the machinery of government in a direction different from what either acting by themselves would have done; but at the same time in a direction partaking of each and formed out of all."[3]

The system by which strategy was formulated in England during the War of the Spanish Succession was a decentralized one in which colonial governors, envoys, and commanders in the field and at sea significantly influenced decisions at home and acted as the eyes and ears of the government, providing the insight, recommendations, and understanding upon which decisions were based. At the same time, the decisions made in London were the products of several committees or boards such as the Board of Trade and the Admiralty, acting on advice from abroad and making recommendations on it to the cabinet. For much of the war, the cabinet itself had a variety of committees which dealt with various problems and prepared matters for cabinet discussion, made detailed arrangements, and supervised the execution of plans approved in cabinet. The various secretaries of state were the key administrative agents for the cabinet, but they were not by themselves the decision makers. The cabinet made its decisions through a consensus of all its members, in the presence of the queen.

Marlborough was both an envoy and a commander in chief in the field as well as a member of the cabinet, but there is no documentary evidence to suggest that he usurped the process of cabinet government under the Crown.[4] At this point in its development, the British cabinet system had not yet developed to the stage in which there was a prime minister who was universally recognized as primus inter pares, although the person who was the closest to being in that position was Lord Godolphin, not Marlborough.

From this one concludes that English strategy was not the product of any one individual, but made by a more abstract and less personal, perhaps less interesting, method. If no single individual expressed fully the concept of English strategy, then, if it is to be found at all, one must search in the interchange of ideas among government officials who represented the process by which decisions were made and through which the armed forces found their command and direction. This was the means by which decisions were made as well as by which the government expressed its policy in the conduct of the war.

The lack of strategic planning documents in a twentieth-century style makes it necessary to construct artificially an outline of England's basic strategic view from disparate sources and varied documents. In order to do this one must look first to the private and public correspondence of all of the various envoys and diplomatic representatives, to the admirals at sea, the generals in the field, to colonial governors, to the Board of Trade, the Admiralty, the orders of the secretaries of state, the scattered notes of officials, and the records of expenditure in the Treasury. Through this one can formulate a composite picture, drawing from a phrase here and a paragraph there while systematically examining a very large range of documents covering a dozen years of warfare.

The Strategic Situation

Looking out at the world in the opening years of the eighteenth century, England saw Europe, tired and financially exhausted by the long wars of 1672–1678 and 1689–1697, but faced with a dilemma created by the death of King Charles II of Spain in 1700. The problem had been foreseen for nearly forty years, but there was no easy solution. In its narrow sense, it was the dynastic question: Who would be the next king of Spain in succession to the childless Charles? If it had been merely a legal question as to who would inherit his titles, it could have been easily settled. But it was not, because the succession of a new king involved the most fundamental considerations in European international politics.

The death of Charles brought an end to the house of Habsburg in Spain. One dynasty had ended, another was needed to take its place. The prince who would succeed to the Spanish throne and who would rule the weak nation with its vast territories around the world, would be in a position that carried little power of its own in Europe. The political and family connections which the new prince brought with him to the throne, however, could profoundly affect the other European nations by bringing the Spanish dominions and trade with Spain and her territories under the direct control of one of the major powers.

The laws of inheritance to the Spanish crown produced two major candidates: King Louis XIV of France and Emperor Leopold I of Austria. A Bourbon claim versus a Habsburg claim. If either should obtain full right to the entire Spanish inheritance, either themselves or through their children, the balance of power as well as the balance of trade would be affected.

Both England and the Dutch Republic were deeply worried, although neither had a claim to the succession. For the Dutch on one hand, the Spanish Netherlands had served as a bulwark of defense against the French, and would remain so if it were controlled by a third power such as an independent Spain or Bavaria. Under French control, it could be the avenue of attack from France as it had so often been in the past. The power which controlled the Spanish Netherlands also controlled the mouth of the Scheldt with its trade entrepots and the port of Dunkirk, which could so easily be used to threaten England and to interrupt Anglo-Dutch sea links in the North Sea. On the other hand, the Dutch and English trade pattern in America, Asia, the Mediterranean, and to Spain herself could be diminished or even cut if Spain became dominated by France. If an independent prince succeeded to the Spanish throne, that trade could be preserved or expanded. As early as 1698, William III had made it clear to France that a Bourbon succession to the entire Spanish monarchy would mean a war with England and the Dutch Republic, as well as with Austria, which opposed that succession because of its own dynastic claim. Between 1698 and 1701 England was actively involved in diplomatic negotiations which sought a solution to the problem.[5] In the negotiations England sought to establish a partition of Spanish territory which would create a balance of political power, thereby assuring the maritime powers that they would not be excluded from any area of trade.

On All Saints' Day 1700, King Charles II of Spain died, and the following day his will was opened in the presence of a large group of nobles. The document declared that no part of the Spanish monarchy was to be divided from the main body. The worldwide interests and possessions of Spain were to be maintained for the next generation of Spaniards. Philip, duke of Anjou, the grandson of Louis XIV, was named in the will as the successor to Charles II as King Philip V of Spain. Failing him, the succession would pass to his younger brother, the duke of Berry, and next to the Habsburg archduke Charles.

It was up to France to decide whether the will of King Charles would be accepted or whether the principles of the previous partition treaties with England and the Dutch would be honored. After the news of the will was received in Paris, the subject was considered in detail by the French government. Opinion on the proper course of action was divided; however, after a full consideration, Louis XIV accepted the will and proclaimed the duke of Anjou as King Philip V of Spain.

England's Objectives

Although France had rejected the Partition Treaty with the English and the Dutch, it would be eighteen months before war was actually declared. Parliamentary and public opinion in England was divided on the issue. A large number of people opposed any war as ruinous to the nation's commerce and believed that England should not enter a war unless she were attacked.[6] Others saw that there was little that England and Holland could do if Parliament insisted on disbanding the army, as the popular sentiment generally demanded after a war.[7] There was a possibility that France would overrun Spain and that England and Holland would fight between themselves for the riches of the Indies. Preventing the French from having "such accession of riches to their Empire, whereby they will be enabled to give laws by sea and land to all Europe," could be achieved by using the English and Dutch navies alone, many argued.[8]

As stadholder as well as king, William's first concern was to prevent the Spanish Netherlands from falling into the hands of France.[9] By the very nature of the events, William saw that, despite opinion at home, Europe would not long remain at peace. He thought that the wisest course of action was to make preparatory agreements with the northern crowns and with as many of the German princes as possible.[10] Backed by Dutch opinion, he encouraged the States-General to begin negotiations in these matters, although he was prevented from doing so in England. The impeachment of several ministers by Parliament in the previous year for their part in the Partition Treaties made it clear that Parliament did not approve of English involvement in these Continental affairs.

It was not until the sudden movement of French troops into the Spanish Netherlands during the night of 5 and 6 February 1701 that English opinion changed to a strong determination to prevent further French encroachment.[11] Secretary of State Sir Charles Hedges made this clear when he told the English representative at The Hague: "You will see by the proceeding of both Houses of Parliament, and especially the Commons that we are awake and sensible of the too great growth of our dangerous neighbor, and are taking vigorous measures for the preservation of our selves, and the peace of Europe."[12]

In order to achieve these aims, Parliament authorized the king to enter into negotiations with other powers in Europe and to conclude the necessary alliances. In June 1701, Marlborough was instructed by the king to undertake these negotiations at The Hague "for the Preservation of the Liberties of Europe, the Property and Peace of England, and for reducing the Exorbitant Power of France."[13] These elements were the basic points upon which England proceeded in her negotiations for an alliance.

On 7 September 1701, the representatives of England, the Dutch Republic, and Emperor Leopold I signed the Treaty of Grand Alliance, "having thought a strict conjunction and alliance between themselves necessary for repelling the greatness of the common danger." The treaty itself outlined the basic issue in its preamble. While objecting to the claim to the Spanish throne that Louis XIV made for his grandson, the Allies deplored the movement of French forces into the Spanish Netherlands, the Duchy of Milan, and the West Indies. Most important, they feared that the succession of Philip in Spain could signal a union between France and Spain which would "within a short time become so formidable to all that they may easily assume to themselves the dominion over all Europe."[14]

The central issue for England was to remove French capacity to dominate Europe. It was not known with certainty whether France had definite plans to expand her territory and control, but English policymakers worked from the reasonable assumption that France under Louis XIV would proceed to direct Spanish affairs as much as possible. Thus, the dynastic connection indicated an imminent expansion of French power in Spain, as well as in Spanish possessions in Italy, the Netherlands, and overseas. When French troops moved into the Netherlands and northern Italy, English statesmen, as well as others in Europe, presumed that France not only could but intended to exploit the dynastic connection created by the accession of Philip V to the Spanish throne. Such a sudden growth of French power would affect England in her most sensitive areas, King William said: "In respect to our Trade, which will soon become precarious in all the valuable Branches of it; in respect to our Peace and Safety at Home, which we cannot hope should long continue, and in respect to that part which England ought to take in the Preservation of the Liberty of Europe."[15] An English diplomat in Switzerland put the issue more bluntly. "Nothing but force or some blow to the French prosperity will make them tractable . . . ," he wrote. "You can have no security but their weakness."[16]

Although peace continued in name, the other nations of Europe were forced to arm themselves and to prepare for war in order to defend themselves from possible attack. The physical security of the Dutch Republic was directly threatened by the French troop movements into the Spanish Netherlands. The connection between the security of Holland and that of England had been long understood, as had the strategic importance of the Channel's far shore for England's defense.[17] There was a clear danger if any enemy obtained unimpeded control of the Continental coast east of the Strait of Dover. Many people remembered that it had been a "Protestant," easterly wind which had allowed William to sail past the English fleet, immobilized by wind and tide in the Thames estuary, and to land at Torbay as recently as 1688.[18]

The methods of William's invasion were well remembered a dozen years later, as were also the reasons for his succession to the throne. The death of Anne's son, the young duke of Gloucester, meant the end of the Stuart dynasty in England and the need to settle the English succession anew.[19] In 1701 Parliament made provision for this in the Act of Settlement. After Anne, the crown was to go to the nearest Protestant heir, the dowager electress Sophia of Hanover. The right of Parliament to regulate the succession had been established in 1689, and it had been tacitly accepted by Louis XIV at the Peace of Ryswick in 1697. The Act of Settlement by Parliament was an expression of English opinion; it was no guarantee against foreign intervention in the English succession. The presence of William's predecessor, the Catholic James II, at the court of Louis XIV was not reassuring in any way. The Prussian representative in London observed at the time that the union of France and Spain presented a direct threat to the Protestant interest in England. Indeed, many believed that the continued growth of Catholic power abroad threatened to destroy Protestantism in England.[20] In Vienna, the English diplomat George Stepney despaired that he could never persuade the emperor to assist wholeheartedly in England's objectives. He feared that Leopold would achieve his own goals secretly through the mediation of the Pope and the Jesuits, "and then leave us to struggle as well as we can for our

Liberties and Religion whenever France or Spain shall join together to impose upon us a Prince of Wales, a Duke of Berry or anybody else."[21]

A few days after the signing of the Treaty of Grand Alliance at The Hague, James II died. Louis XIV immediately proclaimed James's son as King James III of England, Scotland, and Ireland. France's instant and unqualified public recognition of the pretender as king shocked the people of England. William III immediately ordered the English ambassador to return from Paris without taking leave of the French court,[22] and at the same time dismissed the French representative in London. The French recognition of the "pretended Prince of Wales" puzzled Englishmen and left grave doubts as to French intentions. The recognition seemed a direct challenge to Parliament's right to regulate the succession to the throne and to the very principles established by the settlement of the Glorious Revolution. Coming at a time when relations were very tense in Europe and when Louis XIV had just forbidden his subjects to trade with England, it seemed the greatest provocation possible to England short of an outright attack. Amazed by this series of events, an English diplomat at The Hague commented, "Whom God designs to destroy he infatuates first, and makes them do their own business themselves."[23]

At home, Parliament was stirred to take direct action. Early in 1702, the House of Commons passed a resolution which asked King William to insert, in all treaties of alliance with other powers, an article stating that no peace should be made with France until England "shall have reparation for the great indignity offered by the French King, in owning and declaring the pretended Prince of Wales King of England, Scotland and Ireland."[24]

The addition of this article to the Treaty of Grand Alliance, and its subsequent ratification by the emperor and the Dutch Republic,[25] signified recognition by the allies of one of England's major objectives. She sought the acknowledgment by the European powers that the parliamentary title of a Protestant line to the throne of England was superior to the hereditary title of a Catholic line. In seeking this acknowledgment, England was attempting to remove the threat of foreign intervention in an issue which had already been settled satisfactorily at home.

Following the death of William III, the menace of France in Europe was the central issue because it threatened the balance of power. This issue was so important that Queen Anne stressed the point at her accession, when she declared to the Privy Council in March 1702: "I think it proper upon this occasion of my first speaking to you to declare my own opinion of the importance of carrying on all the preparations we are making to oppose the great power of France. And I shall lose no time in giving our Allies all assurances that nothing shall be wanting on my part to pursue the true interest of England, together with theirs for support of the common cause."[26]

When the Allies jointly declared war on France on 15 May 1702, England had limited objectives in view. She did not seek to destroy her enemy. She sought to change her enemy's policies and to limit French capacity for future threats. England's basic aims in entering the war were to secure her own safety, to prevent foreign interference in the revolution settlement, and to secure and maintain her trade abroad. In order to achieve these goals, English statesmen believed that there must be a balance of power in Europe which would hinder anyone nation from interfering with the normal development of another nation. Although this might also bring benefit to others, it would specifically allow England to achieve her primary objectives. The major and immediate threat to obtaining this political situation in Europe was posed by the potential growth of French power through the inheritance of the Spanish throne by the French king's grandson, Philip. The inheritance by a Bourbon prince was not in itself a threat, but military and political trends suggested that the potential danger would become a reality. The Grand Alliance of England, Austria, and the Dutch Republic against France and the Spanish forces of Philip V was based at the outset on the premise that the Spanish inheritance must be divided and that the crowns of Spain and France should never be united. The Dutch and English initially agreed to help Austria obtain a just and reasonable portion of the Spanish inheritance in Italy, but by no means the entire control of Spanish territory.

England's Strategy to Achieve Her Objectives

France was the strongest military power in Europe. In the face of attack, her military objectives were to preserve and defend a geographically contiguous position. Although weaker at sea, she had an army that equaled in size that which the Allies could field against her. In addition, France had the advantage of a single command and logistics system. In her support, France had several key allies. Chief among them was Max Emmanuel, the elector of Bavaria, whom King Charles II had appointed governor in the Spanish Netherlands. His brother, the elector of Cologne and bishop of Liege, allowed the French to occupy key fortresses along the Rhine and Meuse. In addition, there was Victor Amadeus of Savoy, who had defected from an earlier alliance with the English and Dutch in 1696 to rebuild his exhausted principality in northwestern Italy. Then there was Pope Clement XI, whose benevolent neutrality favored King Louis XIV over Emperor Leopold and encouraged the Italian princes in that direction.

On the Continent, the Dutch had the strongest and best-trained army among the allies, numbering nearly one hundred thousand men. The Habsburg army was close to the same size, but less well organized and less well paid. The English had the smallest army of the three. At sea, however, the situation was reversed: the Habsburgs had no capability whatsoever, and the Dutch agreed to provide three-eighths of the naval force, while England would provide five-eighths.

The practical military problem that England faced was how to deal with France's great military strength with the assets available. England's basic strategic idea was to engage the superior strength of France on as many fronts as possible in order to compel her to divide and thus to weaken her forces. Since this was an objective which no single European nation could accomplish alone, the maintenance of an active alliance conducting an offensive war with several armies was the keystone to the strategy. Ideally, England sought to have an alliance which would attack France from the Low Countries, from Germany, and from Italy, and would attack the forces of Philip V in the Iberian peninsula. In this manner, the campaigns in each theater were fundamentally connected in the English understanding of grand strategy for the war.

Several other elements were necessary to maintain this type of war. Allied naval supremacy was essential to support military operations in the peninsula and in Italy, as well as to maintain communications in all areas. In addition, amphibious raids at various coastal points, as well as attempts to create insurrections through dissident Protestant groups, were used to further draw off concentrations of French military forces in the main fronts. In several cases, such as the amphibious landings planned for Picardy and the Boulonnais in northern France, they were designed to create a new focus of operations within a theater. They were an attempt to prevent the type of interminably long and indecisive sieges that occurred at Lille and Tournay.

To complement the military efforts, England planned to use her naval strength to weaken the French and Spanish economies through attacks on the Spanish fleets carrying silver from America back to Spain. In addition, several forms of economic blockade were attempted through a ban on grain.

From 1702 to 1711, each succeeding year saw a renewed attempt to apply the same concept of strategy: encircling of France through an active allied offensive on several fronts, and fighting a war of attrition to achieve limited political objectives. The documents clearly suggest that this broadly sketched concept guided England's employment of her military and naval forces. It also lay behind the use of her financial resources and diplomacy to obtain the proper strategic position, as well as influencing the use of Allied forces to join with England in carrying out the strategy.

To some degree, it was an impractical idea, for it failed to consider the problems of implementation, the varying national goals among the Allies, a French counterstrategy, the impact of events, and the changing political situation during the course of the war. Broadly, the strategy did not aim directly at securing England's basic national objectives by military and naval campaigns. Although some operations were directed toward those objectives, the main goals of military and naval strategy were shaped by the particular situation in Europe at that time and its relationship to the balance of power. Considerable evidence indicates that this dominant theme of strategy remained in use through the very final stages of the war.

When English diplomatic efforts and goals are contrasted with England's actual use and employment of her own resources, it is apparent that the ministry in London was fully aware that its grand strategy could not be carried out by one nation alone.

An ability to move and sustain troops was indispensable to England's direct contribution to the war on the Continent. In particular, the navy had to preserve a safe passage for British troops and munitions of war en route to the Iberian peninsula, the Mediterranean, and along the shorter passage to Flanders. In carrying out these functions, the navy's main role was largely defensive, maintaining a firm basis from which the army could conduct its offensive against France. Yet the navy's offensive role was not neglected. Joining with privateers, the navy attacked French supply lines, engaged French war fleets when they appeared, and often made it imprudent for the French fleet to sail out. In addition, the navy was used on several occasions to present an additional threat to France by attempting to land along the coast and helping to incite revolt among dissidents.

The essential problem, as England understood the war on the Continent, was to provide mobility for the army, to sustain communications in support of her two main armies, and to apply selectively her military, naval, and financial resources to sustain the overall effectiveness of the alliance in attacking France in several theaters.

Problems of Implementation

England's concept of grand strategy was a simple one, but as Clausewitz would later suggest, even the simplest thing in war is difficult. Some of these difficulties were apparent early on, while others appeared in the process of time and in the unfolding of events.

One of the problems which was clear from the outset was the need to augment the Dutch, English, and Austrian armies with men and supplies. The key to doing this was through subsidies to the weaker states and small principalities in order to pay and provide for the troops. This involved England in a complex diplomatic process that resulted in nearly seventy different treaties during the war. More important, it involved large sums of money paid to foreign troops. In most cases, England paid half the costs, sharing the expenditure with the Dutch.

The sources of troops were limited. The Scandinavian countries had supplied many men in earlier wars, but the outbreak of the Great Northern War in 1700 between Sweden, Denmark, Poland, and Russia denied that source to the Grand Alliance. Denmark's early withdrawal from that conflict made her available to provide forces, but as the Great Northern War continued, many small states became apprehensive of their own security and so contributed less than they otherwise might have done. This, of course, was a situation which the French did not fail to exploit.

The key to this system was readily available cash or credit which could be used to buy troops and supplies. This meant that it was necessary to maintain a thriving and expanding economy. In the case of the English and Dutch economies, both of which were highly dependent on overseas trade, a large proportion of naval forces had to be employed in defensive operations that protected trade from enemy attack. This was particularly true for homeward-bound convoys from America and the East Indies, particularly in the final leg of their passage from St. Helena in the South Atlantic. Similarly, in home waters naval vessels provided additional protection for the overseas as well as the coastal trade. A system of patrols covered nearly every section of the British coast. Although there were some coastal fortifications, guards, and garrisons at home, England's defense was primarily at sea.

In the broad terms of grand strategy, England's defensive operations were designed to complement her offensive war against France. Yet in the practical terms of allocating limited naval and military resources, the two requirements created a tension. In order to carry out ambitious attacks against France, it was often necessary to reduce forces used for other purposes.

This tension was particularly apparent in England's American colonies, which showed a growing desire during the war to attack neighboring colonies owned by European enemies. The problem began in the south with expeditions against the Spanish in Florida, and it also occupied the minds of English colonial leaders in the West Indies. The northern colonists were content to remain at peace as long as they could, but when the war fell upon them, they joined in with schemes and expeditions against France. On the North American continent, defenses were scanty and frontier settlements could not withstand a determined assault. Without adequate defenses of their own, the colonists believed that the best way to remove the enemy threat was to destroy or capture the enemy settlements from which attacks were launched. This was the basic rationale behind the attacks on St. Kitts in 1702, Port Royal in 1707 and 1710, and Quebec in 1711. More often than not, the colonists required substantial assistance from home if these projects were to be carried out. The ability of the ministry in London to support these plans depended not only on advance planning but, more importantly, on the state of the war in Europe and the demand for forces in the main theaters of activity. There was little interest in extending English dominion, but there was deep concern for the safety of the existing English plantations and the trade from them.

It was her trade, not dominion, which England sought to enhance. However, the defense of Britain and the protection of English interests abroad were not entirely passive undertakings. The basic interest of England was founded on the economic growth of the nation through a mercantilist system of colonies and overseas trade. In this context, the preservation of trade was also the preservation of economic expansion, but it was not "imperialism" in its nineteenth-century sense.[27]

Strategy in Practice

Strategy is not merely a concept of how to use force for a nation's political goals; it is the actual direction of real force in the context of a situation. In looking at the actual practice of English grand strategy, one can see three major phases corresponding to the periods 1702–1704, 1705–1711, and 1711–1712.

Phase One

During the first period, England tried to position the Alliance to put her full concept of grand strategy into practice. Although the fighting began in the Low Countries, much needed to be done to activate the other theaters. Two major series of campaigns were undertaken. One was designed to pave a way into the Mediterranean in order to encircle France and to launch an offensive against her from the south and southeast. The other was designed to secure Austria and allow her to conduct an offensive against France.

England sought to secure the eastern link of the Alliance by thwarting the attempt of Louis XIV to use Bavaria, with its strong army, as the means to destroy the implementation of England's concept of grand strategy. When English diplomatic enticements failed to bring Bavaria into the Grand Alliance, her defeat at Blenheim in 1704 removed the most dangerous threat to Austria and encouraged both Austria and the German princes to support the Alliance.

Furthermore, England's victory at Blenheim complemented her operations in the Mediterranean,[28] starting with her attempt to seize Cadiz in 1702. When this failed, England sought to attract Portugal into the Alliance in 1703 to provide a gateway to the Mediterranean and a foothold in Iberia. The ensuing capture of Gibraltar and the naval battle of Malaga in 1704 secured the Anglo-Dutch naval squadron's position in the Mediterranean and provided a means to assist Austria by sea. This persuaded Savoy to leave the French fold and join the Alliance at the end of 1704, thus closing the strategic ring around France.

As Sir Julian Corbett pointed out, the English fleet's strategy in the naval campaign of 1703–1704 was to avoid action with the French until England had gained a naval position from which she could not be ejected, thereby forcing France to abandon the struggle for sea communications. England paid a high price to obtain the allies necessary to encircle France.[29] In the case of both Portugal and Savoy, England had to agree to assist them to achieve additional war aims as part of the bargain. Most important, England specifically agreed with Leopold I and Portugal to put Archduke Charles on the throne of Spain. This agreement seemed reasonable in 1703, since it allowed an independent Habsburg prince to wear the crown in Spain and remain free of direct control from either Vienna or Paris. It was hardly to be expected that before the war was over, both the emperor and his eldest son, Joseph, would be dead, leaving the imperial crown to the young Charles.

Phase Two

The period between 1705 and 1711 was marked by England's repeated attempts to orchestrate the forces of the Grand Alliance consistently with her concept of grand strategy. England urged all her allies to attack France simultaneously. English troops continued to fight alongside the Dutch in the Low Countries; imperial armies under princes Eugene and Ludwig of Baden, along with the army of the duke of Savoy, were encouraged to take the field in Germany and Italy. In Portugal, a Portuguese army under Anglo-Dutch command was urged to attack Spain from the west, while an Allied amphibious force landed on the eastern coast. Barcelona and Catalonia were captured in 1705. The following year saw the great battle of Ramillies in Flanders and the capture of the southern Netherlands, plus the allied entry into Madrid. Despite these successes, paralleled by a concerted effort to cut French grain supplies, French forces pushed back the Allies in both Spain and Portugal, defeating the Allied army at Almanaza in 1707. In the same year, the Allies failed in their key effort to seize Toulon by a joint attack using the Allied fleet and the armies under the duke of Savoy and Prince Eugene. Regaining momentum in 1708, the English navy stopped the French attempt to break out from its encirclement and land an army in Scotland under the old pretender. In the Low Countries, Marlborough won victories over the French at Oudenarde and Malplaquet.

At the same time, several important distractions to Austria and the German princes were removed. The Swedish army invaded the Ukraine in 1709, reversing its earlier westward movement into Saxony, which had seemed to threaten a convergence of the Great Northern War and the War of the Spanish Succession. In Hungary, the emperor gained the upper hand in the Kuruc revolt.[30] Neither of these events, however, had the effect England sought: to bring vigorous offensive operations against France from Austria and Germany. Meanwhile, the war in the peninsula was bringing no decisive results as both sides felt the burden of attrition and the drain on their resources.

As the government in London repeatedly attempted unsuccessfully to activate all theaters of the war for simultaneous attacks on France, the English public became increasingly dissatisfied with a wearying series of indecisive campaigns. This was the background to a series of political maneuvers that brought down the government with Godolphin and Marlborough and elevated Robert Harley to power. A general election in the autumn of 1710 gave the Tory party a large majority and caused further ministerial changes.

This sequence of events had several repercussions. The Harley ministry had come to power on a peace platform; publicly it appeared to advocate peace at any price. Harley himself believed that his policy was to reach an expeditious peace through the established war strategy, a peace which would be just to all the Allies. Nevertheless, the Allies viewed England's new leaders with suspicion. Marlborough, as a court favorite, key diplomat, and victorious general, was regarded by many as the cement of the Alliance. Abroad he was often seen as the all-powerful director of England's war effort, not more mundanely and accurately as an important figure in a complex bureaucratic process of decision making. The decline of his political influence and the change in ministry were wrongly interpreted as an abrupt shift in English war policy, other than the result of internal political and court quarrels.

Marlborough's brilliant victories in Flanders overemphasized in men's minds the importance of that theater and obscured the concept that lay behind England's use of her diplomacy, money, men, and ships. Some historians have implied that public opinion determined English war policy. In this understanding, the public debate seems to reflect serious strategic alternatives based on either the army or the navy, Flanders or Spain, Continental or blue-water strategies. The fact is that the Harley ministry immediately returned to the strategy that England had employed from the outset of the war, as the most effective means to force France to the negotiating table. Continuing defeats in Spain, as well as doubts over the ability and willingness of the other Allies to continue to fight, prevented the Harley ministry from putting all its hope in a military solution, but the outline of the strategy which had been pursued for the previous nine years was clearly its first line of approach.

Phase Three

The Harley ministry had been in power for about a year when a key factor in determining England's policy changed, thereby creating the third phase in the execution of the strategy. In April 1711 Emperor Joseph I died after a reign of only six years. In October 1711 Archduke Charles, the Allies' candidate for king of Spain, was elected Joseph's successor as Emperor Charles VI. His election presented a dilemma for English policy and strategy. As emperor, Charles secured Austria's position in the war against France, but if the grand strategy were to be carried to its logical conclusion, the balance of power would be upset by bringing Spain, Italy, and the Indies under direct Austrian control if Charles was also king of Spain. England, the Dutch, and the emperor had all agreed to support Charles's claim as part of the treaty by which Portugal had been brought into the Alliance in 1703.

In this difficult and embarrassing situation, England tried to continue to use her grand strategy as the most effective means to put pressure on France, while at the same time attempting to secure from this position of strength a peace which was advantageous to the Allies but based on a balance of power. Once a cease-fire had been achieved in 1712, England carefully employed her forces in the Mediterranean to prevent the juncture of Spain and the Empire under Emperor Charles VI, while at the same time procuring the special commercial and territorial arrangements England wanted. From the very outset of the war, England had sought to win an independent Spain, unfettered to any major power in Europe, in order to facilitate the growth of English commerce and to secure political independence from any major European power, and she continued to do so in 1711–1712.

The Allies, however, saw "perfidious Albion" renege on what they thought was a long-standing and secure promise to obtain the Spanish throne for Charles. No longer insisting on victory in Spain, the Harley government appeared to be conducting a duplicitous policy by first encouraging the Allies to fight in all theaters, no longer supporting Charles's candidacy in Spain, and then appearing to seek a peace treaty of its own. This suspicion undermined the effectiveness of the Alliance as an instrument of war. Nevertheless, the Harley government continued to try to keep the Alliance working for its strategic purpose, but by 1712 it was clear that England's viewpoint and objectives in employing her concept of grand strategy were different from those of her allies.

Up until that point, the means of strategy could also serve as the ends. It was the means—the grand strategy that England wished to use against France—which formed the basis upon which the Alliance was built and held together. The Allies agreed to defeat France, and each wished to achieve particular dynastic, financial, diplomatic, territorial, or military goals as part of the Alliance. England utilized these desires so that the Alliance might create the conditions under which she could secure her own ultimate ends, but she rarely employed those ultimate ends as the basis for agreement with the other allies.

Throughout the war, England saw herself as the leader of the Alliance, acting jointly with the Dutch as her closest supporters. However, England's view and conduct were based upon her own national viewpoint, not a broader European or Allied concept of affairs. Like others, she saw herself as the center of things, and she showed difficulty in fully appreciating the viewpoints, needs, and ambitions of her allies. Most important, the English ministry, both before and after 1711, failed to understand how its own pursuit of long-standing goals could appear as a threat to the Allies. While the government in London appreciated certain Allied ambitions to the point of using them as incentives to participate in the war, England appeared to assume that her own concept of a European balance of power as the necessary condition for national independence, commercial growth, and international influence was equally compelling and important to the Allies.

In the same way, England failed fully to appreciate the competing goals of the Allies which distracted them from carrying out the English concept of grand strategy. Although these are not unusual problems in relations between nations, they caused growing tension between England and her allies in the War of the Spanish Succession. In many cases, joint Allied achievements satisfied each ally, but the reasons for satisfaction were, of course, often different. The Blenheim campaign, for example, was undertaken by Marlborough and the ministry at the instigation of Germans and Austrians, but its purpose in English eyes was not merely to rescue Austria from attack but to maintain the Alliance and to facilitate further Allied military operations against France in Spain, Italy, and Germany. In the eyes of others, these priorities were reversed.

In initiating the negotiations that resulted in the Treaty of Utrecht in 1713, England saw herself as acting for the Alliance. She sought what she considered a fair and equitable peace for the Allies, while at the same time reaching her own goals. The ministry quite sincerely believed that it was acting responsibly toward the Allies, but in doing so it failed to understand the nature of their varying points of view. In the end, the Allies could do no more than individually acquiesce to a peace which England had created. Utrecht created a balance of power, partitioned the Spanish monarchy between Habsburg and Bourbon, and gave England substantial advantages in trade, as well as territory at Gibraltar, Minorca, St. Kitts, Newfoundland, and Hudson's Bay. It also secured the Protestant succession in England. Thus, through a war of alliance, encirclement, and attrition, England achieved the goals which had been set out more than a dozen years before.

It was during this last phase, however, that domestic politics forced English leaders to stop the momentum of the war and come to grips with a realistic solution in terms of their objectives. The shift of focus from means to ends is never easy in war. By 1712 both France and England had reached the point where they could make a mutual agreement in the light of their own objectives and the calculated high cost of further warfare. While France could maintain Philip V on the Spanish throne at the cost of commercial and territorial concessions, and partition of the Spanish monarchy, the other allies were forced to reconsider what objectives they could achieve in the circumstances. For some of them, the war was unsuccessful and unleashed the intense internal debates which, in such circumstances, commonly inhibit negotiations.[31]

Conclusion

An analysis of the economic, political, diplomatic, naval, and military aspects of the English view and conduct of the War of the Spanish Succession reveals that England consistently attempted to use all her resources to obtain her own, peculiarly English objectives. International diplomacy, finance, and military and naval force were all used in a complementary fashion to achieve these ends. With singularity of purpose, the successive governments which managed the war were motivated by the fact that the nation's strategic problem was to maintain the balance of power in Europe. This was the best practical arrangement through which England could maintain her national security, political independence, and commercial growth. English seamen, soldiers, and diplomats served, within their own spheres, to achieve those ends.

British Grand Strategy in World War I

Michael Howard

'Grand Strategy' in the first half of the Twentieth Century consisted basically in the mobilisation and deployment of national resources of wealth, manpower and industrial capacity, together with those of allied and, where feasible, of neutral powers, for the purpose of achieving the goals of national policy in wartime.[1]

The above definition, in which I attempted to adapt Clausewitz's famous dictum to the age of total war, was prescriptive rather than descriptive. All strategy is in principle teleological; military operations should be planned to achieve the political object for which the war is fought. But it is a principle honored more often in the breach than in the observance. Normally the priorities are reversed. In spite of himself the strategist finds that his plans are being shaped by immediate military and political necessities, which cumulatively shape the object of the war. As the British secretary of state for war, Lord Kitchener, remarked at a critical moment in 1915, "We must make war as we must; not as we should like."[2] Indeed, the British experience in World War I provides an excellent case study of this phenomenon.

Clausewitz was perhaps more helpful when he divided wars into two categories: "limited wars" fought for political objectives, and "total wars" fought with the object of overthrowing the enemy and dictating to him whatever peace terms one wished.[3] The German historian Hans Delbrück was further to refine this concept, dividing strategies into *Ermattungsstrategie* (strategy of attrition) and *Vernichtungsstrategie* (strategy of destruction).[4] Those categories do not necessarily coincide. One can employ a strategy of destruction in pursuit of very limited political objectives: this was what Bismarck did in the wars of 1866, against the Habsburg Empire, and of 1870, against France, when the total defeat of the enemy armies in the field was followed by very moderate peace terms. Alternatively, one can employ a strategy of attrition aimed at destroying the very political existence of the adversary, as did Ulysses S. Grant against the armies of the Confederacy in 1864–1865. But the strategy adopted is always more likely to be dictated rather by the availability of means than by the nature of the ends.

In 1914 no one in the mainland of Europe contemplated a strategy of attrition. There were many reasons, psychological as well as military, why "the spirit of the offensive" was so dominant at the outset of World War I, but not the least was an awareness of the fragility of the social consensus underlying European societies. Everywhere ruling elites confronted restive working classes whose socialist leaders appeared increasingly active and threatening. In his massive study *La Guerre future*, Ivan Bloch had warned that the attrition likely to be produced by modern weapons on the battlefield would lead to economic and social collapse at home,[5] and his prophecy had already been borne out in the case of his own country, Russia, in 1905. Count Alfred von Schlieffen, respected as the greatest authority on war in Germany if not in the world, had written in a famous article in 1909 that "a strategy of attrition cannot be conducted when the support of [armies numbered in] millions demands the allocation of billions [of marks]."[6] Rapid victory was seen as a social imperative as well as a military one: a short war was not so much "an illusion" as an essential.[7]

In any case, the strategic pattern which dominated all military thinking was that established by the experience of war in western Europe during the previous century, as conducted by those two great military masters Napoleon Bonaparte and Helmuth von Moltke the Elder: mobile operations, aimed at securing a rapid and decisive victory over the main enemy forces in the field. It was a concept which shaped the prewar thinking of British military strategists as much as it did that of their Continental contemporaries. Naval thinkers, it is true, drew a different conclusion from the Napoleonic experience, arguing that the imposition of an effective blockade would in itself be decisive by rendering it impossible for the enemy to carry on the war. But in the prewar controversy between this view, as set out by the Admiralty, and the arguments presented by the General Staff of the army—that all strategic effort should be devoted to rapid and effective intervention in those great battles which were likely, as they had in 1870, to decide the outcome of the war—the political authorities pronounced in favor of the latter.[8] It was the general belief in prewar Britain, as it was more generally in prewar Europe, that the conflict would be resolved in a series of bloody and protracted battles in France and Poland to which all force should be committed as quickly as possible. Beyond these it was pointless even to try to peer.

But assuming that those battles were victorious, what political end, in British eyes, were they intended to serve? What were Britain's "war aims"? We must always be careful to distinguish between "war aims," the specific things which belligerent states hope to do if they win the war, and the reasons why they go to war in the first place. In Britain's case, however, there was not much difference between the two. Her motives in entering the war in August 1914 were strictly defensive: to preserve a political status quo in which she was the dominant global power. The assertions made at the time and continued throughout the war, that she was fighting to preserve the British Empire, were not purely rhetorical. The status quo had been increasingly under threat from German ambitions, and Germany's specific target was seen by the British, on the evidence of the German naval building program and domestic propaganda, to be not only Britain herself but the entire Pax Britannica, the system whereby British power dominated the extra-European world; a system which the rising power of Germany would replace.[9]

The only *immediate* and *specific* war aim announced by the British government in 1914 was the restoration of Belgian integrity, an objective which harmonized British strategic interests on the Continent with her claim to be the vindicator of international law. But beyond this lay the more ambitious objective of destroying, if not the power of Germany, then at least its will to power. British public opinion had been increasingly conscious of the threat to international stability posed by the Prussian *picklehaube* and jackboot, by the expansionist nationalism and glorification of force to be found in the works of Heinrich von Treitschke and Friedrich von Bernhardi. The triumph of these militaristic sentiments over the forces of moderation was clear to most Britons even without their knowing the declaration of war aims set out by the German chancellor, Bethmann Hollmeg, in his

"September Program" of 1914.[10] Germany had therefore to be purged of "Prussianism," as a century earlier France had been purged of Bonapartism, and reformed once again as a cooperative member of the Concert of Europe.

Nobody believed in 1914, as they came to believe in World War II, that this would have to be achieved by a process of drastic political and social engineering involving the virtual transformation of Germany society at the hands of her conquerors. In 1914 thinking about the making of peace, as about the conduct of war, was dominated by the experience of the nineteenth century. The defeat of the German armies in the field, and the consequent occupation of German territory, would, it was hoped, bring a chastened German government to the conference table. This government would agree to the restoration of Alsace-Lorraine to France, and would persuade Austria to cede Bosnia, Herzegovina, and perhaps other Slav lands to Serbia. If the German armies had been totally defeated in 1914, this might have happened. We may regret that it did not. At that stage of the war, though the strategy pursued by the British, in association with her allies, was one of overthrow, their war aims were still quite limited. They had no wish to chastize Germany to such an extent that her cooperation could not be relied on if imperial rivalries with France and Russia were to erupt again after the war. It was indeed fear of French and Russian as much as of German ambitions that was to lead Britain, once Turkey was enrolled among her enemies, to formulate plans for territorial acquisitions in the Middle East which would link her African and Asian possessions, giving her assured dominance of the Indian Ocean and its riparian territories.[11]

But in 1914 the question what to do *after* the war hardly came up for consideration. During the first months of the war, the problem was how to avoid defeat as the German armies swept irresistibly over France and Belgium; after that it was how to defeat those armies in the field. The question before the British government was how best it could contribute to this objective, within the framework of its alliance with France and Russia.

In 1914 the small British army, with all its senior staff and commanders, had been sent almost in its entirety to France, where it had played a notable part in staving off disaster. The General Staff appreciation had been correct: the expected great battles had indeed taken place, and without the contribution made by the British Expeditionary Force at the first battle of Ypres they might well have been lost. As it was, those battles remained indecisive. This outcome, or rather the lack of one, strengthened the hand of the traditionalists among British strategic thinkers. If the war was prolonged, as was now clearly going to be the case, then Britain should limit her land commitment as she had in previous wars, and exercise her strength mainly at sea. This would enable her to blockade the Germans, keep open her supply lines to the outside world, build up financial and economic strength with which to help her Continental allies as she had in former coalition wars, and employ her limited military capability flexibly, as her "command of the sea" should enable her to do.

This was probably the view held by the secretary of state for war, Lord Kitchener, Britain's most distinguished soldier, who was effectively military dictator during the early months of the war. Since he set down nothing on paper and discussed his plans neither with his civilian colleagues in the cabinet nor with the General Staff at the War Office, we cannot be sure. Kitchener certainly foresaw the war lasting as much as three years. He broke with all precedent in raising a volunteer army, which would eventually number over a million. It is not, however, clear what he eventually meant to do with it, and he may not have known himself. It would not, in any case, be ready for action for at least two years. The most plausible explanation of Kitchener's intentions has been advanced by David French. According to this view, Kitchener calculated that by the time the British army was reaching its peak of effective strength in two or three years' time, the forces both of their adversaries and of their allies would have fought themselves to a standstill. With their own strength still relatively intact, the British would then be in a position to ensure such peace terms as would best safeguard their own interests, and not emerge fatally weakened to confront their traditional adversaries.[12]

In support of this interpretation it must be remembered that Kitchener was an imperial soldier who had spent his whole career extending and defending the British Empire. As commander in chief in Egypt from 1892 to 1899, his principal long-term adversary had been the French, whose forces he had confronted and faced down at Fashoda in 1899. As commander in chief in India from 1902 to 1909, he had devoted himself to organizing the British and Indian armies in the subcontinent for a war against Russia. Alliance with these powers, in his view, could only be seen as temporary: with the disappearance of the German menace the old rivalries would surely reappear, and the British Empire had to be favorably placed to meet them.

The desire to avoid a major commitment to Continental warfare at the expense of imperial interests had been an element in British strategic thinking since the sixteenth century, and Kitchener was to find much support among his cabinet colleagues in his reluctance to commit the new armies to France any sooner, or in any greater numbers, than was strictly necessary. He and they were in consequence open to suggestions as to where else those armies might profitably be employed. The most attractive suggestion was that brought forward in November 1914 by the first lord of the Admiralty, Winston Churchill, that they should be used as part of an amphibious force to open the Dardanelles, capture Constantinople, and knock Turkey out of the war. Once Churchill had made the proposal, his cabinet colleagues rallied to support it. Such a stroke, they claimed, would bring direct help to a hard-pressed Russian ally by distracting her Turkish adversaries from the Caucasian front and opening her Black Sea ports to Allied supplies. It would attract allies in the Balkans, notably Bulgaria and Greece, thus bringing relief to an even harder-pressed ally, Serbia. And the defeat of Turkey would enable Britain not only to safeguard her position in Egypt by establishing a foothold in Palestine, but to extend her influence in the Arabian peninsula and the Persian Gulf, whose rich oil deposits were already seen as possessing global strategic significance.[13]

The strategic advantages of such an operation appeared indeed to be so attractive that no one worried too much about the tactical difficulties. In consequence, the forces sent to seize the Dardanelles were led and equipped as if for an old-style colonial expedition against a culturally and technologically inferior adversary—rather like the Egyptians in 1882. The result was a disaster. The failure of the Dardanelles campaign, like that of subsequent offensives on the western front, demonstrated all too clearly that strategic objectives, however well conceived, can be achieved only by tactical success. Strategic planning has to be determined not only by political goals but by the effectiveness of the military instruments available. In any case, the strategic rationale of the Dardanelles campaign was itself highly questionable. The elimination of Turkey from the war would not have affected the basic strength of Germany. The neutral Balkan powers were more influenced in their policy by the course of events on the eastern front, where Russia and Germany were contending for mastery, than by anything that happened in the Mediterranean. And even if the Allies had established themselves in the Balkans, their logistical problems would have placed them at a decisive disadvantage in relation to the Germans—as they were indeed to find a few months later when they landed a force at Salonica. The decisive theaters, the only ones where the war could be lost or won, were the eastern and western fronts.

In 1915, whatever British strategists may have intended, the eastern front was the major theater because the Germans had decided to make it so. During the course of that year the German armies in the east inflicted such drastic defeats on Russia that her Western allies began to doubt her capacity, and even more the will of her government, to carry on the war at all. It was the need to relieve the pressure in the east that compelled the French and the British armies to continue their offensive on the western front. There was no longer any expectation of a strategic breakthrough leading to a major decision: the object now was to pin down the German forces and exhaust them. It was a strategy determined by the French High Command, and one into which Kitchener allowed himself to be drawn only very unwillingly. But if he did not do so, he feared, not only the Russians but even the French (who had already suffered over a million casualties) might be tempted to make peace. It was at this stage that the truth broke in on him that one has to make war, not as one would like to, but as one must.

Almost imperceptibly, therefore, during 1915 British strategy became transformed. In 1914 British forces had been committed to the Continent to carry out a strategy of overthrow, to help destroy the enemy in a few decisive battles. Now they remained to pursue a strategy of attrition—a strategy in which, as French strength gradually ebbed, the British Empire would have to play an ever-increasing part.[14]

In such a strategy the British could succeed only by wearing down enemy resources and willpower at a faster rate than their own. Such a strategy, moreover, was dictated also by tactical necessity. The hugely increased effectiveness of firepower, the virtual disappearance of mobility from the battlefield, and the high concentration of defensive strength along a limited front all meant that destruction of enemy forces, and occupation of ground, could be achieved only by artillery fire. This was an arm which the industrially powerful states of Western Europe were well equipped to provide, and in 1915 they began tooling up their entire economies, increasingly with the support of the United States, to provide it.

This resulted, in Britain, in the major strategic debate of the war. The issue was not, as it was later depicted in wartime memoirs, between "Westerners" and "Easterners," those who wished to concentrate on the western front and those who hoped to find an "indirect approach" via the Mediterranean against Germany's allies.[15] The failure of the Dardanelles campaign had effectively ruled out the latter option. It was to be revived again two years later, when in the autumn of 1917 the long misery of Passchendaele had discouraged even the greatest enthusiasts for the western front. In 1915 the argument lay between, on the one hand, those members of the cabinet who believed that Britain should revert to her traditional strategy of limiting her commitment to Continental warfare and making her principal contribution through her industrial, financial, and naval power; and, on the other, those who advocated the total and immediate mobilization of all resources, whatever the financial and social cost, to achieve a rapid and total victory.

The principal advocates of the first course were the ministers responsible for the economic management of the war: Stephen McKenna, the chancellor of the Exchequer, and Walter Runciman, president of the Board of Trade. Both belonged to the conservative wing of the governing Liberal party, a coalition of moderates and radicals presided over with increasing difficulty by Prime Minister Herbert Asquith. They found their support among all who believed in sound finance and those landed, traditional elements in society who dreaded the social costs of the total commitment advocated by the "out and outers." The latter consisted of the military themselves; the upwardly mobile business classes who were increasingly filling the seats in the House of Commons; the mass-circulation press; and the radical politicians. The latter were led by David Lloyd George, who was rapidly becoming the dominant figure in Asquith's cabinet and was to replace him as prime minister a year later; and by Winston Churchill, who, though temporarily in disgrace as a result of the Dardanelles fiasco, was soon to be recalled to office, first as minister for munitions and then as secretary of state for war. The first group was haunted by the prospect of national bankruptcy; the second, by that of losing the war.

The debate came to a head in the autumn of 1915 over the proposal to introduce "compulsory national service," or conscription, as a less libertarian society would have called it.[16] After the massive response of the first few months of the war, voluntary enlistment was no longer producing men in the numbers needed for the expansion of the army and the replacement of casualties, which the military authorities foresaw as necessary to sustain the massive offensive they were now planning. Conscription was therefore demanded by the radicals, to produce an army which would once more make possible a strategy of overthrow. It was opposed by the traditionalists on the grounds that it would eviscerate Britain's industrial and economic capacity to carry on a strategy of attrition. The radicals carried the day. The first phase of conscription was introduced at the end of the year. Simultaneously the Allied military leaders agreed, at a conference at Chantilly, that they would launch coordinated offensives on both the western and the eastern fronts, to force the Germans to sue for peace before the Russians—and not inconceivably the French—did so first. The British contribution to this joint offensive, when Kitchener's new armies would be committed in their entirety, was to be the battle of the Somme. Even if the attack did not result in a breakthrough, as General Sir Douglas Haig continued to hope that it would, it was expected to inflict such punishment on the German armies that their leaders would be forced to sue for peace.

But on the Somme, as at the Dardanelles, tactical failure was to ruin the grand-strategic concept. The task was anyhow made more difficult by the German seizure of the initiative with their attack on Verdun in February 1916; a stroke intended by the German High Command to bring on a struggle of attrition to bleed the French armies to death, and one which substantially reduced the forces available for the joint Allied offensive projected for July. But more to the point was the inadequate training and inexperience of the new British armies at every level, the clumsy tactics they adopted, the technical failures of a too-rapidly-expanded supporting artillery, and, above all, the belief that heavy fire-support and meticulous planning could take the place of elementary battlefield skills.

The failure of the assault on the Somme drew Britain into that long ordeal of attrition which both parties in the strategic dispute had hoped to avoid. Since she was taking the offensive at a time when the development of firepower favored the defensive, the scales were tilted against her. The Russian Revolution and the near collapse of the French armies in the spring of 1917 made matters yet worse. Further, British access to the industrial and agrarian resources of North America was threatened, not only by the success of the German submarine campaign, but by the exhaustion of British credits in the United States. In the spring of 1917, barely six months after the "Big Push" had been launched on the Somme with such high hopes of victory, British war leaders were having to contemplate the possibility of defeat. The entry of the United States into the war in April was to save them, not so much because it provided new armies—well over a year was to pass before these became effective—as because it provided them with unlimited credit, and with the cooperation of the U.S. Navy in defeating the German submarines.

The events of 1916 revealed another grisly development in the nature of industrialized warfare. "Attrition" was no longer a question of wearing down the strength of the enemy armies in the field. Their casualties could be made good so long as there were men available to replace, and weapons to equip, them. The real target, the "center of gravity," as Clausewitz termed it, was thus no longer the armies themselves: it was the capacity, and even more the will, of the nation to go on supplying those armies with men and materiel. It was, in other words, the cohesion of the enemy society, that element which had before 1914 appeared so alarmingly fragile but which had since proved so unexpectedly tough. The concept of Vernichtungsstrategie had acquired a new and deadlier meaning.

The object of the war was not total on both sides. Nothing could have shown this more clearly than the part played by the German government, in April 1917, in facilitating the passage of Lenin and his Bolshevik colleagues from Switzerland to Russia to hasten the collapse of the Provisional Government. Peace was to be gained at the cost of destroying a social structure with whose stability that of the whole of European society was connected. On the other hand, the British government, from 1917 onward, initiated a sustained campaign of political warfare in support of independence for the dissident Slav minorities of the Habsburg monarchy—a state whose multinational structure had preserved stability in this most volatile of regions for centuries past and whose disappearance was to leave Eastern Europe a prey to the warring ambitions of Germany and Russia.[17] And finally, in November 1918 President Woodrow Wilson was to grant peace terms to the Germans only on condition that they followed the example of the Russians and overthrew the monarchy which, with all its faults, had provided a focus of loyalty for the nation and whose disappearance was to leave the field open to far more sinister forces. From 1917 onward the belligerent powers of Europe, with the full support of populations inflamed by wartime propaganda, were set on a course of mutual social and political as well as military destruction. Britain was no exception.

Could peace have been made before the war entered this last, disastrous phase? The short answer is yes—but only on Germany's terms. Until the end of 1917 the German armies had been consistently victorious. On the eastern front Russia had been defeated and the Balkans overrun. On the western front the French and British had been fought to a standstill: incapable of further offensives, they awaited with apprehension the assault which the Germans were preparing for the following spring. The American entry into the war had been nullified by the success of the German submarine campaign, in which the tide was only just beginning to turn. Until the end of 1917 no peace would have been possible which did not leave Germany master of the Continent.

But with the Americans behind them, neither the British nor the French would accept a German peace. By the early summer of 1918 it was clear that the German armies in the west could not win the war before their navy lost it. By then social disintegration in Germany was far advanced, and by the autumn the flood of American arms and equipment into Western Europe made it possible for the Allies to dictate their own terms.

What conclusion can be drawn from all this? Certainly Britain's original grand-strategic objective, of reducing the power of Germany without destroying or permanently antagonizing her, had been wise. Equally, the most effective way of doing this, had it been possible, would have been to defeat her armies in the field in a strategy of overthrow. The strategy of attrition into which the British found themselves forced was mutual suicide, as everyone had feared, although the process took longer than expected, and the extent to which British strength had been eroded militarily, economically, and socially did not become clear for another generation.[18] But the alternative would not have been some skillful and painless "strategy of indirect approach." It would have been to stand aloof while Germany destroyed first the Russian and then the French armies, then to make such peace as was possible before German submarines starved Britain into submission; or, at best, to wait for the United States to enter the war. But before 1917 it appeared highly unlikely that the United States would enter the war, or be very effective if she did.

With hindsight we can see that in industrial societies there is no way in which attrition of armies can be separated from the attrition, moral as well as social, of peoples. After the war a few enterprising strategic thinkers in Britain began a quest for the philosophers' stone of a "knock-out blow." Some hoped to find it in the bomber aircraft, others in the tank. But the conventional wisdom was that another war would be very much like the last, and the sensible thing therefore would be to avoid it.[19] It was a belief that was to dominate British statecraft for the next twenty years.

Churchill and Coalition Strategy in World War II

Eliot A. Cohen

It is often held that Winston Churchill made his great contribution to the winning of World War II through the indomitable spirit he displayed, and the glorious rhetoric that he used to inspire Britons—and all free peoples—during a long and bloody struggle. But was he a strategist and, more to the point, a good one? On this question both his colleagues and the historians have disagreed with one another. In a diary published after the war, for example, one chronically jaundiced military adviser quoted the Australian prime minister Robert Menzies approvingly: "Only Churchill's magnificent and courageous leadership compensated for his deplorable strategic sense."[1] Other military advisers and observers have dissented strongly from this view,[2] which has found its most damning formulation in the diaries of the chief of the Imperial General Staff, Alan Brooke. Today, the undeniably heroic quality of Churchill's leadership, coupled with the well-known eccentricities of his life during the war, have left the image of a brilliant and irascible figure, indispensable, no doubt, but a "bull in a china shop," not a calculating and canny strategist. It requires no small effort to evaluate Churchill in the latter light, filtering out the dazzle of the mythic hero, and putting in proportion the petulant, bullying stubbornness that drove his generals to distraction. And one of the best ways to discern the nature of Churchill's warcraft comes through a study of his handling of the problem of coalition warfare in 1940–1945.

Intellectual Preparation: *The World Crisis* and *Marlborough*

Unique among democratic statesmen, Churchill assumed the prime ministership in 1940 having spent some three decades as a mature man pondering the problems of high command and experiencing them first hand, in both peace and war. Churchill drew heavily on that experience in his administration of the British government in World War II. So too did many of his contemporaries, who discussed their present dilemmas and prospects with reference to the problems and solutions of the war of 1914–1918.

It is of particular interest to see how Churchill reflected on the lessons of 1914–1918 for the conduct of coalition war in his four-volume account of the war, *The World Crisis*. Churchill turned to history, including the history of his own life, not merely for amusement or self-justification, but for instruction. He had a remarkable capability for learning from his own mistakes. Although unrepentant in defending the Dardanelles operation of 1915, for example, he conceded his own errors in forcing a military move only lukewarmly supported by senior commanders, and was careful to devise in 1940 a system that would prevent a similar reoccurrence in the future.[3]

In his reflections upon the events of World War I, Churchill placed the greatest emphasis on the coalitional nature of the struggle. Many years later he told his secretary that one need read only *Liaison 1914* and *Prelude to Victory* by Edward Spears—beautifully written accounts focused on alliance politics—to understand that war.[4] A natural Clausewitzian, Churchill asserted the primacy of politics, and particularly the politics of coalitions, in the conduct of that war: "At the summit true politics and strategy are one. The maneuver which brings an ally into the field is as serviceable as that which wins a great battle. The maneuver which gains an important strategic point may be less valuable than that which placates or overawes a dangerous neutral."[5] Of the three errors which doomed the Germans, the two most important were, in his view, mistakes of coalition strategy. By invading Belgium the Germans brought Britain into the war from the outset; by declaring unrestricted submarine warfare they then brought the United States into the war.[6] Similarly, Churchill argued, the Germans squandered their strategic opportunities, which lay in striking mortal blows against the weaker members of the Allied coalition, in search of political as well as economic and more narrowly military effects.

Thus, in 1916 the Germans should have shunned the western front for Russia and Eastern Europe; in 1918 they should have knocked Italy out of the war rather than throw all their remaining forces against the western front in a reckless all-or-nothing bid for European supremacy.[7]

By the same token, Churchill thought that the Allies had repeatedly endangered their cause by failing to integrate their national military efforts. War is a whole, Clausewitz has written, and "in war more than in any other subject we must begin by looking at the nature of the whole."[8] Churchill believed that in 1914–1918, "the war problem, which was all one, was tugged at from many different and disconnected standpoints. War, which knows no rigid divisions between French, Russian and British Allies, between Land, Sea, and Air, between gaining victories and alliances, between supplies and fighting men, between propaganda and machinery, which is, in fact, simply the sum of all forces and pressures operative at a given period, was dealt with piecemeal. And years of cruel teaching were necessary before even imperfect unifications of study, thought, command and action were achieved."[9] Simple stupidity or mulishness did not account for this culpable fragmentation of strategy, in Churchill's view. Some of the fault lay with politicians and generals too slow to learn the imperatives of war; some too lay with innate and near-irreconcilable national differences, for "every great nation in time of crisis has its own way of doing things."[10]

Churchill thought the dilemma a terrible one. The spirit of compromise and patience, so valuable to harmony and progress in peace, would, in war, "only mean that soldiers are shot because their leaders in Council and camp are unable to resolve." In war, however, "clear leadership, violent action, rigid decisions" are not only necessary but even merciful. In war things left alone do not improve by themselves or go away but "explode with shattering detonation."[11] Fortunately, Churchill thought, the flexibility of democratic government allowed states to develop the vigorous executive power required for a single nation to conduct war. Only the greatest efforts by statesmen, however, could secure unity and celerity in a coalition of states. Thus Churchill, in some respects a severe critic of Marshal Ferdinand Foch, applauded his understanding of the need for Allied unity, both on and off the battlefield. Unlike Philippe Pétain, he "ranged the strategic necessities of the Allied armies in their true order. Of these the first beyond compare was the union of the French and British armies."[12]

Two key characteristics of coalition warfare, then, come to the fore in *The World Crisis*. These are (1) the importance of strategic maneuvers aimed at rending apart an opposing coalition and enlarging one's

own; and (2) the contradiction between the unity of war as a phenomenon and the tendency of states—and lesser organizations—to
approach it in a compartmented fashion. In the 1930s Churchill explored these twin aspects of coalition strategy further in his six-
volume biography of his great ancestor, the duke of Marlborough
(1650–1722).[13] A characteristic blend of pessimism concerning the
"feebleness and selfish shortcomings of a numerous coalition" and
optimism about the possibilities of inspired statecraft pervades the
book.[14] "The history of all coalitions," Churchill wrote, "is a tale of
the reciprocal complaints of allies."[15] He found in the War of the
Spanish Succession a parallel with the war just fought, and with a war
yet to come:

> It was a war of the circumference against the centre.
> When we reflect upon the selfish aims, the jealousies
> and shortcomings of the Allies, upon their many natural
> divergent interests, upon the difficulties of procuring
> common and timely agreement upon any single neces-
> sary measure, upon the weariness moral and physical
> which drags down all prolonged human effort . . . we
> cannot regard it as strange that Louis XIV should so long
> have sustained his motto, 'Nec pluribus impar.' Lying in
> his central station with complete control of the greatest
> nation of the world in one of its most remarkable ebulli-
> tions, with the power to plan far in advance, to strike
> now in this quarter, now in that, and above all with the
> certainty of implicit obedience, it is little wonder how
> well and how long he fought. The marvel is that any
> force could have been found in that unequipped civiliza-
> tion of Europe to withstand, still less to subdue him.[16]

Again, Churchill depicted the inevitability of national divergences in
coalition war, and in particular the tendency of weak or mortally
threatened states to try to shift burdens to stronger allies, to choose
caution over audacity, and equivocation over decision.[17] "The armies
of a coalition cannot be handled like those of a single state," he wrote.
It was central to Marlborough's greatness, in Churchill's mind, that he
understood and accepted that fact while struggling to forge a coherent
strategy from the "intrigues, cross-purposes, and half-measures of a
vast unwieldy coalition trying to make war."[18]

The language Churchill used to analyze strategy displayed remarkable continuity during the thirty years between the beginning of the
first world war and the end of the second. In *The World Crisis* and
Marlborough, the reader finds the same themes and diction of the

great state papers of 1940–1945. Churchill invariably picks out the two or three "supreme facts" governing a situation, and lays particular emphasis on the "time factor." Whether speaking of coalition politics at the turn of the eighteenth century or the future strategy of an embattled Britain in 1941, he attempts to assay the "true proportion" of various interests, and speaks of the task of "weaving an ally into the texture of the war."[19] Where his approach appears to have changed over the period 1914–1941, it is in the direction of appreciating ever more thoroughly the problematic character of coalition warfare. Even in his public pronouncements during World War II he continually reminded his audiences that each ally would "naturally see [each] theatre [of war] from a different angle and in a somewhat different relation."[20] Churchill's understanding of the difficulty of achieving unity in the conduct of war did not lead him to a simple pessimism— to the contrary, it spurred him on to greater efforts—but it did color the whole of his strategic approach to the war.

The Tiers of Coalition Strategy

The Equals

Churchill faced a four-tier problem in the conduct of coalition strategy during World War II. In the first and by far the most important of these tiers, he worked with his two great partners in the Grand Alliance, the United States and the Soviet Union. He summed up the importance of these relationships in his speech of 15 February 1942, given after one of the greatest British disasters of the war, the fall of Singapore:

> How do matters stand now? Taking it all in all, are our chances of survival better or are they worse than in August 1941? How is it with the British Empire or Commonwealth of Nations? Are we up or down? . . .
>
> When I survey and compute the power of the United States and its vast resources and feel that they are now in it with us, with the British Commonwealth of Nations all together, however long it lasts, till death or victory, I cannot believe there is any other fact in the whole world which can compare with that. That is what I have dreamed of, aimed at, and worked for, and now it has come to pass.
>
> But there is another fact, in some ways more immediately effective. The Russian armies have not been conquered or destroyed.

> *Here then, are two tremendous fundamental facts*
> *which will in the end dominate the world situation and*
> *make victory possible in a form never possible before.*[21]

In this, as in other wartime speeches, Churchill revealed with remarkable candor the premises of his strategic calculus.

Churchill did not, however, believe that the "two tremendous fundamental facts" of American participation in the war and Soviet perseverance in the land battles in the east had appeared as the result of some natural law of international politics. An effective Grand Alliance could not come into being without sustained British effort, he believed. Particularly once the immediate threat of invasion had passed in the winter of 1940, Churchill had as his first priority the development of that alliance.

After the war Churchill described his emotions upon first learning of the Japanese attack on Pearl Harbor:

> At this very moment I knew the United States was in the war, up to the neck and in to the death. So we had won after all! Yes, after Dunkirk; after the fall of France; after the horrible episode of Oran; after the threat of invasion . . . we had won the war. England would live; Britain would live; the Commonwealth of Nations and the Empire would live. How long the war would last or in what fashion it would end, no man could tell, nor did I at this moment care. . . . We should not be wiped out. Our history would not come to an end. We might not even have to die as individuals. Hitler's fate was sealed. Mussolini's fate was sealed. As for the Japanese they would be ground to powder. . . . Many disasters, immeasurable cost and tribulation lay ahead, but there was no more doubt about the end.
>
> Silly people—and there were many, not only in enemy countries—might discount the force of the United States. Some said they were soft, others that they would never be united. They would fool around at a distance. They would never come to grips. They would never stand blood-letting. . . . But I had studied the American Civil War, fought out to the last desperate inch. American blood flowed in my veins. I thought of a remark which Edward Grey had made to me more than thirty years before—that the United States is like 'a gigantic boiler. Once the fire is lighted under it there is no limit to the power it can generate.' Being saturated and satiated with

emotion and sensation, I went to bed and slept the sleep of the saved and thankful.[22]

Two things are remarkable here: first, Churchill's confidence in victory, which despite the catastrophes of early 1942 did not waver, unlike the less sanguine opinion of his advisers;[23] second, his implied admission that until 7 December 1941 he *had* considered it possible that the war could end in defeat of a peculiarly horrible kind.

Churchill made his greatest contribution to British victory in World War II through his management of relations with the United States, from the period of growing American support of Great Britain in 1940 and 1941 to the forging of the Allies' basic strategy in 1942 and 1943, to the final defeat of the Axis powers in 1944 and 1945. From the first, Churchill appreciated more fully than did any of his colleagues the import of American participation on Britain's side. Indeed, if Churchill had any grand strategy for victory in 1940 and 1941, it consisted chiefly in arranging the entrance of the United States into open hostilities with Germany in partnership with Great Britain.[24] Of course, Churchill was not alone in thinking alliance with the United States highly desirable, even imperative. But he did surpass both civilian and military colleagues in the War Cabinet and military staffs in his unremitting concentration on this task, and in his appreciation of American military and industrial potential.

From the first, Churchill was prepared to make considerable concessions to the Americans in order to bring them ever closer to the war. In August 1940 his War Cabinet colleagues expressed reservations about President Franklin D. Roosevelt's offer to swap fifty obsolete American destroyers for British bases in North America. Churchill reminded them that "if the proposal went through, the United States would have made a long step towards coming into the war on our side. To sell destroyers to a belligerent nation [is] certainly not a neutral act."[25] Similarly, when the Cabinet noted with alarm the withering of British financial resources, Churchill told his colleagues that "if the military position should unexpectedly deteriorate, we should have to pledge everything that we had for the sake of victory, giving the United States, if necessary, a lien on any and every part of British industry."[26] Indeed, throughout the war, but particularly in the period before 1941, Churchill examined every major strategic decision with one question in the forefront of his mind: How will this affect the attitude of the United States, and its eventual entry into the war?[27]

Churchill believed that the United States had enormous and indeed

decisive war potential.[28] Many other members of the British government did not share this view, or at least hold it with anything like
Churchill's confidence. More recent memories of American military
performance in World War I—during which the United States had
raised large armies but often found itself forced to equip them with
European weapons—did not augur well. In 1941 and 1942 Britain's
military leaders thought their own country "soft" and their ground
forces not quite equal to the Germans, but the Americans seemed if
anything less ready for or suited to war.[29] In January 1942 Sir John Dill
wrote from Washington to his successor as chief of the Imperial General
Staff, Alan Brooke, "Never have I seen a country so utterly unprepared
for war and so soft."[30] On the whole, British military experts thought
poorly of the doctrine and efficiency of American ground and air forces,
the former consisting of hastily raised conscript armies, the latter
wedded to unrealistic doctrines of precision daytime bombing.[31]

Moreover, the tremendous military productive capacity of the
United States did not become fully apparent to foreign observers until
the last three years of the war:[32] In other words, in 1940 the United
States produced less than half the volume of munitions of Great Britain, in 1941 about two thirds as much, in 1942 twice as much, in 1943
more than three times as much, and in 1944 almost four times as
much. Whereas in 1941 Great Britain had already reached 59 percent
of her maximum military output, the United States had achieved only
11 percent of hers.[33] In the end, 13.4 million American munitions
workers produced four times as much as 7.8 million of their British
counterparts.[34] As the war progressed, Britain became ever more dependent on the United States for war materiel. Before Pearl Harbor
American munitions supplies to Great Britain were, in the words of
the official history, "negligible." In 1942, however, a tenth of British
munitions came from the United States, and in 1943–1944 well over a
fourth did.[35] In certain key areas—tanks, landing craft, and transport
aircraft in particular—American production made up between 40 and
50 percent of British requirements.[36]

Three forces shaped Britain's influence in the Grand Alliance: the
human and material burdens that it carried, the weight of forces it
could bring to bear on the enemy, and above all the skill of its leaders
in persuading their Allied counterparts to accept British views. Table
1 reveals how quickly the relative war-industrial *weights* of the
United States and Great Britain changed in the course of 1943. The
relative *military* efforts—the forces they could bring to bear against
the Axis powers—changed equally sharply in 1944. Until the spring
of 1944 the British had something like a quarter or a third more divi-

Table 1

Volume of Combat Munitions Production of the Major Belligerents in
Terms of Annual Expenditure
($ Million 1944 U.S. Munitions Prices)

Country	1935-1939	1940	1941	1942	1943	1944
United States	1.5	1.5	4.5	20.0	38.0	42.0
United Kingdom	2.5	3.5	6.5	9.0	11.0	11.0
USSR	8.0	5.0	8.5	11.5	14.0	16.0
Germany	12.0	6.0	6.0	8.5	13.5	17.0

sions in fighting contact with the enemy than did the United States: by January 1945, however, the Americans had forces at least 60 percent larger than those of Great Britain in contact with the enemy.[37] Similarly, the American air offensive against Europe, negligible during 1942 and limited until the end of 1943, did not mature until 1944.[38]

Britain's bargaining position vis-à-vis the United States, then, deteriorated markedly throughout the war, and in several dimensions. Nor had Churchill begun the war with a terribly strong hand: in 1940 and 1941 Britain nearly bankrupted itself, even as its forces were driven from the European continent in one debacle after another. As a result of repeated British defeats in France, Greece, and North Africa, Americans entertained the gravest doubts about British military capabilities and strategy.[39] Tension pervaded the Anglo-American relationship in many areas, and particularly in the Far East. American officials, including President Roosevelt, had few compunctions about pressing the British hard on colonial and economic questions such as postwar arrangements for air transport.[40]

It is therefore all the more remarkable that Churchill and his advisers were able to have their way on virtually every major strategic decision until the end of June 1944, when he was forced to accede to American wishes concerning the invasion of southern France. The invasion of North Africa in 1942, followed by the Italian invasion in 1943 and the invasion of France in 1944, represented the unfolding of a Churchillian strategic design not too different from that laid out in his memoranda prepared for the first Washington conference in 1942. Even in the very last stages of the war Churchill retained considerable freedom of action: he conducted a largely successful intervention in the Greek civil war in December 1944 despite American opposition, which included a temporary cessation of all American naval support to British forces in the area. Churchill deliberately drew an excessively rosy picture of the Anglo-American "special relationship" in his World War II memoirs, in large part because of his belief that Britain's future as a great power rested on the American connection. Despite the corrective judgments of recent historians that the Anglo-American alliance included much in the way of conflict and competition, the special relationship did—and does—exist. Cooperation in foreign policy, defense planning, and intelligence gathering and assessment continues to this day, in no small measure as a result of the foundations laid by Churchill. Judged by the standards of previous alliances, rather than an abstract notion of harmony, the Anglo-American alliance was—and remains—an extraordinary achievement.[41]

Churchill built the Anglo-American alliance in three ways. His unique correspondence with Franklin Roosevelt, and the rapport they established in repeated conferences, is well known. In the course of the war he sent some 950 messages and received 750 from the president, for an average of one exchange between the two every three days. In almost all cases Churchill drafted the messages himself, which indicates the importance he assigned to this correspondence. "No lover," he once remarked, "ever studied every whim of his mistress as I did those of President Roosevelt."[42] One of the shrewdest observers of the two men notes that their relationship rested on an "easy intimacy," but "neither of them forgot for one instant what he was and represented or what the other was and represented. Actually, their relationship was maintained to the end on the highest professional level. They were two men in the same line of business—politico-military leadership on a global scale—and theirs was a very limited field and the few who achieve it seldom have opportunities for getting together with fellow craftsmen in the same trade to compare notes and talk shop. They appraised each other through the practiced eyes of professionals and from this appraisal resulted a degree of admiration and sympathetic understanding of each other's professional problems that lesser craftsmen could not have achieved."[43] A rare sense of comradeship pervaded this relationship, but it too must be placed in proportion. Neither Churchill nor Roosevelt thought that it alone would make relations between the two allies smooth and easy, or eliminate the fundamental differences of interest and viewpoint between them.

This pursuit of alliance through personal relationships extended to other individuals as well. Churchill made a point of cultivating General Dwight Eisenhower, for example, and, in the months before the invasion of Europe, sharing a meal with him as frequently as twice a week.[44] Similarly, from the first Churchill understood the importance of Harry Hopkins as Roosevelt's emissary and confidant and worked—successfully—to make of him a close friend. As he did with British commanders, he would invite these two Americans in particular to spend weekends at his country residence, Chequers, combining sociable relaxation with strategic debate. In no case did the intimacy thereby established prevent disputes from arising, but it did create an atmosphere of goodwill in which they could be resolved.

Churchill operated at two other levels besides the strictly personal in dealing with the Americans. First, from the very beginning of the war he addressed the American public and its elites. The heroic speeches of 1940 and 1941 had American as well as English audiences—a fact Churchill understood and appreciated. Thus, it was no accident that he warned on 18 June 1940 that "if we fail, then the whole world, *including the United States,* including all that we have known and cared for, will sink into the abyss of a new Dark Age, made more sinister, and perhaps more protracted, by the lights of perverted science."[45] No less important were his brilliant and enthusiastically received speeches given in the United States, particularly his address to Congress on 19 May 1943 and his speech at Harvard University on 6 September of that year.[46] Churchill's position as Britain's warlord became virtually unchallengeable by war's end because of his personal popularity: widespread American admiration served him almost as well.

Churchill also built the Anglo-American alliance through the institutionalization of cooperation. Despite his reputation as an unpredictable maverick, Churchill understood the organizational dimension of strategy and applied himself to it. The creation of the Combined Chiefs of Staff in 1942—a move supported by Churchill over the reservations of the British Chiefs of Staff—was perhaps the most important of these organizations, but there were others. The committees that pooled British and American transport and munitions production, the combined intelligence effort at Bletchley Park and in the field, and the various combined commands for northwest Europe, the Mediterranean, and Southeast Asia—all bear a Churchillian imprint.

With the other great ally, the Soviet Union, Churchill never achieved, nor did he desire, a relationship anywhere near as close as that with the United States. He did maintain a personal correspondence with Stalin, although it never achieved the length or intimacy of that with Roosevelt.[47] Throughout the war Churchill strove to strike a balance in his relationship with the Soviet Union. His declaration in 1939 that, the Molotov-Ribbentrop Pact notwithstanding, the Soviet Union had a community of interests with the Western powers,[48] had been borne out by events. His core beliefs—which shaped most of his dealings with the Soviets through the years 1941–1945—remained those uttered on the day of the German invasion of Russia. On the one hand: "the Nazi régime is indistinguishable from the worst features of Communism. . . . No one has been a more consistent opponent of Communism than I have for the last twenty-five years. I will unsay no word that I have spoken about it." On the other: "But all this fades away before the spectacle which is now unfolding. The past with its crimes, its follies and its tragedies, flashes away. . . . Any man or state who fights on against Nazidom will have our aid. Any man or state who marches with Hitler is our foe. . . . The Russian danger is therefore our danger, and the danger of the United States, just as the cause of any Russian fighting for his hearth and home is the cause of free men and free peoples in every quarter of the globe."[49]

Churchill, like his colleagues, periodically indulged himself in illusions about the nature of the Communist regime in the Soviet Union, and about Stalin himself. On the whole, however, four fundamental and enduring beliefs shaped his view throughout the war: first, the Soviet Union would prove more resilient than most of his military advisers initially expected, and one of the chief forces in the ultimate victory over Hitler; second, it was imperative, nonetheless, to keep the supply line to the Soviet Union open even at very high costs, in order to sustain the Soviet military effort and to preclude any possibility of a despairing compromise peace; third, it made little sense to appease Soviet rudeness or bluster, and above all to give in to Soviet insistence on an invasion of Western Europe in 1942 (and later, 1943); fourth, communism in general and the Soviet Union in particular would remain a hostile, if not mortally threatening, force after the war. In this blend of views he differed from the Americans and some of his colleagues such as Lord Beaverbrook, who favored an early second front and had an excessively benign view of the Soviet future; he disagreed too with his military advisers, who initially underestimated Soviet military capabilities and doubted the worth of strenuous efforts to resupply the Red Army.

In the summer of 1941 most observers, including Churchill, expected the Red Army to be defeated, although Churchill rated its chances rather higher than did the Chiefs of Staff and the British intelligence community.[50] Thereafter, he insisted on the imperative need to run supplies to the Soviet Union, even at horrendous costs. The British official history notes that in retrospect "our loss in parting with [these supplies] was greater than the Russians' gain in receiving them,"[51] but the scale was hardly miserly.[52] Churchill told his colleagues that losses of up to 50 percent on convoys to the Soviet Union would have to be regarded as acceptable.[53] It was "our duty," he told the cabinet, "to fight these convoys through, whatever the cost."[54] And the cost was high, not only in terms of men drowned and ships and cargoes lost, but also of scarce resources diverted from the defense of the Far East. The 450 aircraft shipped to the Soviet Union by September 1941 could have done some service in defense of Singapore, a painful fact which Churchill acknowledged and accepted. "If the Malay Peninsula has been starved for the sake of Libya and Russia," he told Attlee at the end of December 1941, "no one is more responsible than I, and I would do exactly the same again."[55]

Churchill went further. He reportedly urged upon the Chiefs of Staff consideration of Operation Jupiter, a landing in northern Norway—a scheme which they repeatedly, and in retrospect quite properly, rejected as impracticable. This proposal had as its chief purpose the securing of routes to the Soviet Union, which lay exposed to German air and naval forces poised to pounce on the northern convoys.[56] Operation Jupiter may well have had its origins in Stalin's first wartime message to Churchill (18 July 1941), which called for the opening of two fronts against the Germans, one in northern France, the other in the Arctic.[57] His insistence on the consideration of this and similar operations gained impetus from the reluctance of the Chiefs of Staff to approve any offensive measure beyond the bombardment of Germany on the one hand, and his acute awareness of the burden being borne by the Soviets on the other. His explosive minute to the Chiefs of Staff of 8 April 1943 bears quotation in this context. General Eisenhower had warned that the presence of two more German divisions would rule out the invasion of Sicily, to which Churchill declared:

> If the presence of two German divisions is held to be decisive against any operation of an offensive or amphibious character open to the million men now in North Africa, it is difficult to see how the war can be carried on. Months of preparation, sea power and air power in abundance, and yet two German divisions are sufficient to knock it all on the head. . . . I trust the Chiefs of Staff will not accept these pusillanimous and defeatist doc-

trines, from whoever they come. . . . I regard the matter as serious in the last degree. We have told the Russians that they cannot have their supplies by the Northern convoy for the sake of 'Husky', and now 'Husky' is to be abandoned if there are two German divisions (strength unspecified) in the neighbourhood. What Stalin would think of this, when he has 185 German divisions on his front I cannot imagine.[58]

For this reason as well Churchill was more reluctant than his Chiefs of Staff to abandon the idea of a second front in France in 1943.[59] The figure of 185 German divisions (a slight underestimate, in fact) would return again and again in Churchill's strategic arguments with the Chiefs of Staff and the Americans.

At the same time, Churchill's suspicions about the Soviets never entirely disappeared, and indeed grew throughout the war. He did not think it necessary, for example, to yield to Soviet demands for a guarantee of their pre–June 1941 borders, which included the Baltic states overrun by the USSR in the wake of the Molotov-Ribbentrop Pact.[60] As early as October 1942 he told Anthony Eden that "it would be a measureless disaster if Russian barbarism overlaid the culture and independence of the ancient States of Europe."[61] In March 1943 he expressed his fear that the United State would abandon Europe after the war, and declared to a confidant, "I do not want to be left alone in Europe with the Bear."[62] In August of that year he passed to Roosevelt a long and grim account of the Katyn Forest massacre, an event about which both had known since at least April.[63]

Churchill, acutely aware of the precariousness of British strength and conscious of the surge of Soviet armies into Europe, strove to make the best deal he could with Stalin. This accounts for his efforts to persuade the Polish government in exile to accept Soviet territorial demands in return for a measure of independence in the postwar period, as well as for the famous influence-sharing arrangements with Stalin in October 1944. He could count on some American support, but only in a limited measure, as his disagreement with Roosevelt over how to react to the Warsaw uprising in August 1944 indicates. Despite Churchill's repeated pleas that the Anglo-Americans brush aside Soviet objections and drop supplies to the insurgents, Roosevelt—motivated in part by the need he foresaw for Soviet bases in the war against Japan—refused to cooperate.[64] Similarly, his urging of an Anglo-American liberation of Berlin and Vienna fell on deaf ears.

Churchill opposed Soviet policy in many respects, but he could not afford to break with Stalin during the war. This resolve stemmed chiefly from his sense of Britain's weakness, Russia's strength, and America's ambivalence. But it may have had roots also in his fundamental assessment of the nature of communism, and its ultimate ability to create an effective, warlike society. The Nazi-occupied, war-devastated societies of Europe would, he feared, remain for a time vulnerable to communism, even if not occupied by the Red Army. "The penalties of defeat are frightful. After the blinding flash of catastrophe, the stunning blow, the gaping wounds, there comes an onset of the diseases of defeat. The central principle of a nation's life is broken, and all healthy normal control vanishes."[65] Yet even at the dawn of the Cold War Churchill retained more hope for the future of the contest with the Soviet Union than he had had in the 1930s regarding the struggle with Hitler. The West's lead in the manufacture of nuclear weapons, and the creation of an Anglo-American alliance based on a resolve not to repeat the errors of 1933–1939, sustained his confidence, even in the wake of such titanic events as the loss of China to the Communists.[66] As he repeatedly told audiences and intimates, the Soviets feared the West's friendship more than its enmity, and he did not believe that it would prove possible for the Russians to maintain their vast empire indefinitely.[67]

The Empire

Students of coalition strategy in World War II justly concentrate their attention on the relationship among the Big Three—the United States, Great Britain, and the Soviet Union. Yet other coalition partners required adroit handling and consideration by Churchill, and of these none were more important than the members of the British Empire and Commonwealth, particularly India and the self-governing white Dominions, Australia, New Zealand, Canada, and South Africa. Their ultimate contributions to the war effort were considerable and varied. Canada, for example, produced 8 percent of the British Commonwealth's total supplies, although it was an exception.[68] More important was the manpower supplied by the Dominions and their strategic depth. When the British government laid its plans for the creation of fifty-five-division army, it expected twenty-one of those divisions to come from the empire.[69] Moreover, Commonwealth and imperial divisions bore the brunt of the battle in particular areas, especially in Greece in 1941 (where two Australian and one New Zealand division provided the bulk of the forces).

The empire gave Great Britain a superb geostrategic position. A global archipelago of bases provided, as in the past two centuries, control of key crossroads—the exits from the Mediterranean, the tip of Africa, the approaches to the Indian Ocean. Even better, these bases existed in depth: a fleet based at Singapore could (and did) fall back on Trincomalee or even East Africa. Were Gibraltar to be neutralized, British forces could still operate from West African bases to dominate the western approaches to the Mediterranean. The empire also served as a gigantic and (except in early 1942) a secure rear area, particularly for training aircrew on a titanic scale. Just as important, the existence of the empire created in Britain a sense—exaggerated perhaps—of power and resilience beyond that to be expected from a cluster of relatively small islands on the margins of the Eurasian land mass. On the other hand, the empire also created many difficulties for British strategy, particularly in the Far East. The exposed positions of Australia and, to a lesser extent, New Zealand meant that British planners had to worry about how to defend those dominions from Japanese attack. Domestic turbulence in India—rising to something approaching insurrection in 1942—threatened the mainspring of British strength in Asia.

Of more immediate concern to Churchill and his advisers in 1940 and 1941, however, was the incorporation of the empire into the war effort. It should be noted that this did not proceed without difficulty. Although Canada, Australia, and New Zealand entered the war in 1939 almost immediately, in South Africa only a narrow vote and the statesmanship of Churchill's close friend Jan Smuts brought that country into the war swiftly and fully on Britain's side. Furthermore, the Dominions, and Australia in particular, did not simply raise forces and hand them over to the British High Command to be used as Churchill and the Chiefs of Staff saw fit. In 1941 Australia pressed for the creation of an independent Australian army corps in the Middle East, which involved the complicated (and, to the British commander in the Middle East, unwelcome) withdrawal of one division from the besieged fortress of Tobruk.[70] Australian influence was consolidated by the appointment of the Australian lieutenant general Thomas Blamey as deputy commander in chief Middle East—a position which did not, however, prevent him from reporting directly to his government. Tensions with Australia—bad enough by the end of 1941—worsened sharply when the Japanese entered the war.[71] The Australian government successfully pressured the British High Command to send home Australian divisions fighting in the Middle East. Similar, though lesser, difficulties arose with respect to the other dominions.[72] In some cases, the alliance with the United States exacerbated the problem of managing the imperial coalition. Canada's contiguity with the United States and integration with her war economy weakened one tie; Australia's dependence on the United States for military protection after December 1941 nearly eliminated another; and sharp American criticism of Britain's handling of India strained yet a third.[73]

Churchill navigated his way through these problems with some care. Although he quite agreed with General Claude Auchinleck concerning the folly of withdrawing Australian troops from Tobruk, for example, he insisted that the British government would have to acquiesce: "Allowances must be made for a government with a majority of one faced by a bitter Opposition, parts of which at least are isolationist in outlook. It is imperative that no public dispute should arise between Great Britain and Australia. All personal feelings must therefore be subordinated to the appearance of unity. Trouble has largely arisen through our not having any British infantry divisions in the various actions, thus leading the world and Australia to suppose that we are fighting our battles with Dominion troops only."[74] Churchill therefore went out of his way to assure the Australian government that "at whatever cost your orders about your own troops will be obeyed."[75] This latter statement reflected a principle that Churchill applied throughout the war: he might try to persuade, cajole, or even bully Dominion governments into appropriate action, but in the final analysis he would not—and perhaps could not—order their forces into action against their will.

Churchill made a point of keeping Dominion governments well posted on the broad progress of the war, sending their heads the Chiefs of Staff *Weekly Résumé* (by surface mail, however), as well as daily telegrams on the progress of operations, summaries of cabinet papers of particular importance, and personal communications to the prime ministers of the Dominions drafted by Churchill himself.[76] On a number of occasions Dominion prime ministers visited Great Britain and were invited to take part in War Cabinet discussions. Moreover, in his reconstruction of the War Cabinet in February 1942, Churchill gave the deputy prime minister, Clement Attlee, the additional job of secretary of state for the Dominions—a gesture of more symbolic than real import.[77]

From the very first, however, Churchill rejected the nation of creating an imperial War Cabinet composed of prime ministers or their representatives. He laid out his reasoning in a speech to Parliament on 17 January 1942: "To hear some people talk, however, one would think that the way to win the war is to make sure that every Power contributing armed forces and every branch of these armed forces is represented on all the councils and organisations which have to be set up, and that everybody is fully consulted before anything is done. That is, in fact, the most sure way to lose a war."[78] Churchill created some institutions—most notably the Pacific War Council—as a forum for imperial debates about strategy, but these remained feeble and, in the end, unimportant. Strategy for the war would be forged by the president and the prime minister, aided by the Combined Chiefs of Staff. Furthermore, Churchill did not believe in sharing important war secrets, particularly the details of forthcoming operations, with his colleagues in the Dominions, with the exception of his valued counselor, General Jan Smuts. When Mackenzie King, in late May 1944, asked for a general indication of when D-Day might occur, Churchill told him it might occur as late as 21 June, the backup date in the event of a postponement of the landings on 5 and 6 June. The Canadian official historian records that this remark "might be uncharitably defined as an exercise in the art of how to deceive a Dominion prime minister without actually lying."[79]

Imperial considerations thus played a curiously ambiguous role in Churchill's coalition leadership. This held true of Churchill's strategic conceptions as well. He burned to avenge Britain's humiliation in the Far East, and, in what was probably his least inspired strategic conception, urged upon the Chiefs of Staff impracticable schemes for an early amphibious counteroffensive to drive the Japanese back from the outposts they had won in the disastrous winter and spring of 1942. On the other hand, he had starved Singapore of reinforcements before the Japanese onslaught, and shared his military chiefs' view that the Americans had to be weaned from their temptation to fight the war in the Pacific first. He would do what was necessary to maintain imperial support for the war, he accepted that imperial defense must remain the first charge on British resources, and he had some sympathy for the domestic political problems of the Dominions. He subordinated imperial considerations, however, to the larger management of the Grand Alliance, and to the winning of the war against the chief enemy, Germany.

Lesser Allies and Neutrals

Churchill's management of relations with the lesser allies and neutrals has a number of points of interest. Of the allies, the most important and the most prickly was Free France, under the leadership of Charles de Gaulle.[80] Here, as in the case of the empire, we see understanding coupled with a strong belief that all relations must be subordinate to the broader interests of the Grand Alliance. This frequently led to clashes with Anthony Eden and the Foreign Office, in which Churchill would side with the Americans against De Gaulle.[81]

For the most part, Churchill treated the governments in exile with consideration and courtesy. His sympathy for the small nations of Europe, defeated and enslaved by the Nazis, took its color from the vivid historical images which dominated his worldview.[82] He was an early advocate of building up foreign units in the British armed forces and opposed efforts to take petty advantage of the weakness of governments in exile,[83] but he did not believe in extending to the smaller allies any greater a decision-making role than that of the Dominions— if anything, even less. In some cases, most notably that of Poland, he showed himself willing to apply considerable pressure in order to force concessions necessary to preserve the postwar peace—and, in this case, he hoped, the existence of a free Polish state.

Churchill could, as the Polish negotiations suggest, take a hard and at times brutal line with the leaders of exiled governments.[84] But this was tempered by a consideration for the occupied peoples, an ambivalence that emerged in several forced and painful military choices. Perhaps the best-known episode of this kind concerns the April 1944 debate over pre-invasion bombing operations in France, which were likely to kill thousands of French civilians. Churchill stubbornly insisted (with the support of virtually all of the War Cabinet) that "there was a limit to the slaughter and resulting anger it [the bombing] would arouse among Frenchmen beyond which we could not go."[85] Only after prolonged argument and the scrutiny of the results of preliminary bombing operations did he and the War Cabinet concur with the wishes of the military planners.[86]

Churchill's attitude to genuinely neutral or nonbelligerent powers was considerably less sympathetic, save in one or two special cases (Switzerland most notably). Three strands of thought are particularly noticeable here. First, Churchill did not think Great Britain should allow itself to be constrained in the conduct of the war by international law or an excessive sensitivity to neutrality. In 1939 he concluded a memorandum arguing for the mining of Norwegian waters in the teeth of Norwegian objections with the following observations: "The final tribunal is our own conscience. We are fighting to reestablish the reign of law and to protect the liberties of small countries. Our defeat would mean an age of barbaric violence, and would be fatal, not only to ourselves, but to the independent life of every small country in Europe. . . . Small nations must not tie our hands when we are fighting for their rights and freedom. . . . Humanity, rather than legality must be our guide."[87] Moreover, Churchill thought such violations of neutrality not only permissible, but in some cases highly desirable, for he thought it imperative to expand the war against Germany in every way. If this meant forcing unwilling neutrals into the war, so be it. In arguing for the seizure of four Swedish destroyers in June 1940, Churchill countered the arguments of those who thought that such an action might provoke a German invasion of Sweden: "A decision on this question should be taken on a broad basis. It might well be that all Europe, including Spain and Sweden, would fall under German control in the near future. But it might well be to our advantage that the Germans should have to hold down all these intelligent and freedom-loving people; the task of thus holding down all Europe should prove beyond even the strength of the Gestapo, provided England could retain her liberty."[88] Similar considerations animated his belief in aid to Greece and Yugoslavia in 1941, and his constant support for the stirring up of insurrection in occupied Europe.

This second aspect of Churchill's attitude to the neutrals, his belief in the necessity of expanding the war at all costs, culminated in his ardent and ultimately fruitless pursuit of an alliance with Turkey until near the very end of the war. Such an expansion of the Grand Alliance seemed to have much to offer: it would open yet another front to make the Germans "bleed and burn"; it would allow greater scope for the application of Allied air power against Rumanian oil, which supplied a third or more of German consumption; it would open a direct route to the Soviet Union; and it would increase the range of strategic threats that the Allies could pose in the Balkans. The cautious Turks allowed themselves to be wooed with arms and training missions, but in the end declined British advances until near the close of the conflict.

The third strand of Churchill's strategy vis-à-vis the neutrals lay in his willingness to adopt and place faith in strategies of preemptive intervention. The campaigns of World War II, as of any global war, raged on the margins of, and sometimes in, neutral or nonbelligerent states whose geographical location lent them an importance well above any forces they could bring to bear. Churchill believed strongly in the importance of swift, if risky, interventions in order to beat the enemy to the possession of such areas. In some cases—Norway in 1940, Thailand in 1941, the Dodecanese in 1943—these operations were either aborted or countered by an equally swift-moving enemy. In others, however, and most notably in the case of Iraq and Syria in 1941, such attacks proved both cheap and successful. Indeed, had such intervention not taken place, the results could have been grim.[89]

Churchill's willingness to endorse such operations required considerable courage. They often faced opposition from military commanders on the scene, for they inevitably drew on makeshift task forces stripped away from already exiguous units. These preemptive attacks required adroit political timing and preparation as well, with a regard not only for the local situation but for the larger regional and even global repercussions. Churchill understood such operations as maneuvers of a particularly delicate yet profitable kind, not nearly as bloody as the slaughter of a set-piece battle but no less useful in the larger scheme of things.

The Enemy

Similar reasoning shaped Churchill's attitude toward the opposing coalition. To Germany and Japan he could offer no more than unconditional surrender: but in the treatment of the lesser Axis powers, most notably Italy, he saw the scope for a strategic maneuver on the grand scale. From the first, Churchill saw the possibility of an Italian collapse, and thought that the Allies should prepare themselves to take full advantage of it.[90] In this he did not, at first, have the support either of the British Chiefs of Staff nor, most certainly, of their counterparts in the United States. That Sicily would serve as a useful follow-on to the North African campaign, all could agree; an Italian campaign, however, faced opposition, particularly from the Americans, who feared a continued draining of scarce resources into the Mediterranean campaign.

The British Chiefs of Staff soon saw the merits of Churchill's position, and joined with him to argue for the prosecution of an Italian campaign. The aim was not, as some have alleged, to break into Fortress Europe through a back door, or (at least for Churchill) to avoid entirely Operation Overlord, the invasion of France. Rather, the purpose was to force the maximum dispersion and effort upon the Germans in two ways. First, the Germans would have to open up a new land front directly opposite the Anglo-Americans: this would have an effect even if the Germans could hold the Allies to a line across the peninsula or even push them back a bit.[91] Second, the removal of Italian forces from the war, particularly in Yugoslavia, would force the Germans to replace them with their own forces. The German units in the Balkans would, moreover, have to cope with opposing guerrillas reinforced by British air and naval assets operating from southern Italy. This said, it must be confessed that the Italian campaign came, in 1944, to exercise an undue fascination on the British Chiefs of Staff and their prime minister, who began to see possibilities for a decisive breakthrough that probably did not exist.

Churchill's distinctive contribution to Mediterranean strategy lay partly in conception, but more in emphasis—particularly in his insistence on the importance of speedy operations to take advantage of an Italian collapse, and in his vigorous support for aid to Tito's Partisans.[92] Success in Italy required, in Churchill's view, a willingness to "deal with any Non Fascist Italian Government which can deliver the goods," even if such a government included the monarchist or military elements.[93] In the end, this strategy met with only mixed success. The Germans proved themselves operational masters of the art of preemptive occupation, as they had in Norway in 1940 and in France and North Africa in 1942, and were to show themselves again in Hungary in 1944. For a variety of reasons—some logistic, some embedded in the cautious operational styles of the British and American armed forces—the Allies found themselves incapable of capitalizing on an opening which their intelligence organizations had foreseen and their political leaders, Churchill in particular, had understood. Churchill, no doubt, often concocted operational schemes that ignored the practical constraints of moving and supplying armies. On the other hand, both American and British general staffs were wedded to rigid and inflexible planning, which by itself would have proved incapable of seizing opportunities and which was, in fact, excessive. Nor did Churchill always err on the side of too much daring, as his support for a larger landing on D-Day than previously planned (five divisions rather than three) suggests.

The Italian campaign did, in fact, serve the purpose Churchill intended, that of draining German resources and avoiding an intolerable period of a year or more in which the Russians would bear the entire brunt of the land war against Hitler. In July 1943 the Germans deployed some twelve divisions in their Southern and Southeastern Commands, which included Italy and the Balkans: by year's end they deployed over thirty.[94] Air operations from southern Italian bases had begun to strike at Germany from the south, and the campaign drained German resources without, in the end, compromising the coup de grace to be delivered at Normandy.[95]

The Strategy of Balance

Churchill's approach to strategy in general, and coalition strategy in particular, had its share of flaws. But the standard students of history must use is not that of omniscience and prescience, but of prudence and sense, recognizing fully the crushing pressures of war leadership. Measured by that standard, Churchill's achievement in World War II was and remains extraordinary. Lord Ismay declared in his memoirs that May and June 1940 might have been Churchill's finest hour, but "the fourteen months which elapsed between the German attack on Russia and the Battle of Alamein [were] his greatest achievement. In the one case he steeled a nation to defy defeat; in the other, he laid the foundation of Allied victory." The basic design of the war was his; so too was the alliance structure that, through the medium of unique institutions such as the Combined Chiefs of Staff and the system of Allied supreme commands, successfully waged it. Churchill's assessment of the psychology of his great ancestor might apply to him: "But Marlborough felt the greatest compulsion that can come to anyone—the responsibility of proprietorship. It had become his war. He was the hub of the wheel. He was bound to function. He had made the treaties. . . . He alone knew the path which would lead them out of their tangles and tribulations, and he was bound to force or trick them to salvation if he could."[96] And it must be recognized that the Allies could have lost the war, if simply by errors which would have led to stalemate. An attempt to invade France in 1943, before the Luftwaffe had suffered crippling losses, and while the Germans still disposed large enough reserves to reinforce speedily in the west, would most probably have failed. If it had, one may doubt whether a second attempt would have been made: the psychological blow to the British alone would have been crushing, and the repercussions for the Soviets perhaps equally devastating. Had Churchill not established intimate bonds with the United States, and in particular with Roosevelt, the Anglo-Saxon powers might have failed to pool their shipping and their productive resources nearly so efficiently as they did. And even so, logistical considerations constrained Allied strategy in all theaters of war to the very end.

The fertility of Churchill's mind, the vividness of his speech, and the vigor of his prose have misled some into thinking of him an impetuous, unstable, and volatile statesman, heroic, no doubt, but hardly a strategist in the proper sense. Yet this view misconceives both Churchill and the nature of strategy itself. Throughout his career Churchill was, in fact, remarkably consistent in his core beliefs.[97] And during the world war his fundamental strategic conceptions remained remarkably true to those laid out in his memoranda on future strategy prepared for the first Washington Conference in December 1941.

Soldiers and historians sometimes think of strategic planning as an exercise in creating a blueprint for victory. During World War II the bomber components of the air forces of both the United States and the United Kingdom adhered to this view; so too, in large part, did the chief

of staff of the United States Army, George Marshall.[98] Conscious of the
enormous technical problems involved in creating forces and planning
operations on the one hand, and confident in their possession of war-
winning weapons and techniques on the other, they thought it at once
possible and necessary to reach for victory in a linear fashion.

Churchill understood but rejected this conception of war. "A operation of war cannot be thought out like building a bridge; certainty is
not demanded, but genius, improvisation, and energy of mind must
have their parts."[99] More than once he told his advisers of his deep
mistrust "of these cut and dried calculations which showed infallibly
how the war would be won."[100] In his autobiography he warned
readers, "Never, never, never believe any war will be smooth and easy,
or that anyone who embarks on the strange voyage can measure the
tides and hurricanes he will encounter."[101] In the fall of 1940 he told
the War Cabinet, "The question might be asked, 'How are we to win
the war?' This question was frequently posed in the years 1914–1918,
but not even those at the centre of things could have possibly given a
reply as late as August of the last year of the war."[102] And as late as the
spring of 1943, he cautioned the Congress of the United States that
"war is full of mysteries and surprises. A false step, a wrong direction,
an error in strategy, discord or lassitude among the Allies, might soon
give the common enemy power to confront us with new and hideous
facts."[103]

Churchill's conception of strategy emerged from this conception of
war.[104] If the ruling metaphor in the minds of some who think about
strategy is the blueprint, Churchill conceived of war as a kind of
picture, and the forging of strategy as a kind of art.[105] The strategist,
like the painter, searches for "proportion or relation" and must adhere
to it: "There must be that all-embracing view which presents the
beginning and the end, the whole and each part, as one instantaneous
impression retentively and untiringly held in the mind."[106] At the
core rest the key facts or themes: "If only four or five main features are
seized and truly recorded, these by themselves will carry a lot of ill-
success or half success."[107] In Churchill's view, then, strategy was
neither a matter of building a machine to narrow tolerances and an
exact design, nor a chaotic welter of unconnected and opportunistic
decisions. The strategist, like the painter, needed a theme and a grip
on the fundamental facts of the situation: as a war, like a painting,
developed, new refinements, different emphases, or changed methods should appear, all governed, however, by a single perspective or
approach.

Churchill wrote of Marlborough's contemporary, Halifax, that "a love of moderation and sense of the practical seemed in him to emerge in bold rather than tepid courses. He could strike as hard for compromise as most leaders for victory."[108] He took this approach himself:

> I thought we ought to have conquered the Irish and then given them Home Rule: that we ought to have starved out the Germans, and then revictualled their country; and that after smashing the General Strike we should have met the grievances of the miners. I always get into trouble because so few people take this line. . . . It is all the fault of the human brain being made in two lobes, only one of which does any thinking, so that we are all right-handed or left-handed; whereas if we were properly constructed we should use our right and left hands with equal force and skill according to circumstances. As it is, those who can win a war well can rarely make a good peace, and those who could make a good peace would never have won the war.[109]

Churchill's conduct of coalition strategy in World War II reflected this vivid-hued harmony. From the first, the main themes—the subject of the picture—were clear, and he never deviated from them: the primacy of the American alliance, the importance of keeping Russia in the war, the task of dismembering the opposing alliance while extending one's own. Proportion governed throughout: the Dominions must be treated with consideration and care, but not allowed to distort strategic decisions or the apparatus that would make them. The Soviet Union must be supported vigorously, but unreasonable demands rebuffed and her long-term objectives suspected. The liberties of small states must be safeguarded, but an overly meticulous concern for their rights should not jeopardize the larger cause at stake.

Strategy is not merely a matter of straightforward arithmetic calculation, a linear process akin to that of building a house.[110] The arts of persuasion and negotiation, of matching jobs with men to fill them, conceiving new organizations and building them, maintaining morale and spurring effort, of discerning first principles and knowing when to apply them, come in as well. This is true of all war, but even more so of grand-coalition wars such as that waged by the Great Britain between 1939 and 1945. Churchill's genius for strategy in World War II may have been both "massive and uneven,"[111] as the official historian puts it, but genius it most certainly was.

The Grand Strategy of the Continental Powers

The Grand Strategy of the Roman Empire

Arther Ferrill

Edward Gibbon said that the Roman Empire "comprehended the fairest part of the earth and the most civilized portion of mankind. The frontiers of that empire were guarded by ancient renown and disciplined valor."[1] That is generally true, although some modern analysts have complained that the frontiers were guarded so well that there was no central reserve, which probably should have been stationed in northern Italy, ready to move anywhere it was needed.[2] In short, a rapid deployment force of the sort created by President Jimmy Carter.

In what has aptly been called this century's best book on Roman history by a nonspecialist, Edward Luttwak has offered a highly compelling survey of the grand strategy of the Roman Empire in a book with exactly that title.[3] He has argued that the emperors consciously followed a grand strategy that can be reconstructed today, with a little help from archeology and the study of fortifications, but mainly on the basis of troop placements on the Roman frontiers.[4] Luttwak believes that imperial grand strategy passed through several distinct stages. The first he describes as a flexible one in which the frontiers were not clearly drawn but, insofar as they did exist, were defended by a combination of Roman legions and client kings, or allied satellite states. This ill-defined grand strategy gradually evolved, by the second century, into the famous rigid-frontier defense system, described by Luttwak as "preclusive security" and characterized in its most famous form by Hadrian's Wall in England.[5]

The perimeter line of defense around the empire was nearly six thousand miles in circumference, and in the second century it was defended solely by Roman troops, since there were no longer any client kingdoms or buffer states. Almost all of them had been absorbed directly into the majesty of the Roman Empire. Unlike Alexander's empire, which was about as large and was conquered virtually overnight, the Roman Empire spread gradually for centuries like a fungus over the ancient Mediterranean. Luttwak strongly criticizes the Roman system of preclusive security, implying that it represented a Maginot line mentality and was deficient for the lack of a central reserve. In the event of an attack on one point of the vast frontier, legions could be moved there along well-maintained roads from other border posts, but that then left part of the defensive perimeter undefended.[6]

A war on two fronts was, therefore, a great danger to the military integrity of the empire, and civil war an even greater one. In the third century both calamities occurred. In the chaotic fifty-year period from 235 to 284, there were more than twenty emperors, only two of whom died a natural death. More than one hundred aspirants for the purple tried to seize all or parts of the empire, as legion fought legion in a great, intermittent civil war that left the frontiers largely undefended. Barbarians poured in from all directions, and one province, Dacia, was permanently lost. Others were ravaged, and the empire nearly collapsed before the great emperor Diocletian restored it, politically and militarily, at the end of the third century.[7]

Finally, the emperors of the fourth and fifth centuries adopted a new grand strategy, and in the most famous part of Luttwak's book he analyzes the new system, which he calls defense in depth. Constantine the Great created a major new army, a central mobile striking force, based heavily on strong cavalry contingents, bivouacked near wherever the emperor happened to be.[8] This field army could move quickly to critical points on the now thinly defended late imperial frontiers—thinly defended because the manpower needs for supporting the central reserve no longer permitted deployment of large numbers of troops on the distant perimeter of defense.

Ironically, although Luttwak's review of defense-in-depth is the most respected part of his book, he actually does not deal with the fourth and fifth centuries of Roman imperial history—the subtitle of his book on Roman grand strategy is, significantly, "From the First Century A.D. to the Third." As we shall see, this title gave Luttwak the splendid opportunity to praise defense-in-depth as a grand strategy on a theoretical basis while ignoring the days of its intensive implementation in the late Roman Empire when the entire western half of the ancient civilized world was lost to barbarism, and defense in depth proved totally ineffective.

Although I do not share Luttwak's views about Roman grand strategy, his book was a seminal one, and it has largely set the framework of subsequent discussions of the topic. Much has been written about his strongly stated argument that Roman grand strategy was a defensive plan consciously applied by the emperors, and many professional Roman historians have doubted this, arguing that it simply emerged slowly on an ad hoc basis.[9] Obviously, some emperors, like some American presidents, paid little attention to legions and frontiers. Nero is the classic example, but Roman dynastic history abounds with scoundrels who cared not at all for frontier problems, preferring the pleasures of Rome to the struggle with barbarians. Sometimes great strategic episodes occurred under such emperors, but that did not make them grand strategists. The grand strategy of Caligula, Nero, Commodus, or Elagabalus would be about as futile a subject as the grand strategy of President Millard Fillmore (even though Commodore Matthew Perry opened relations with Japan in his administration).

Yet many emperors were actively concerned with the defense of the empire. Although they worked within the framework of an inherited defensive policy, one that went back even beyond Augustus to the days of the late Republic, they were generally familiar with the military needs of the empire on a conscious level, and some of them made some significant modifications in Roman grand strategy that were consistent with the overall policy. In addition to Augustus, the Flavians, Trajan, Hadrian, Marcus Aurelius, Septimius Severus, Diocletian, Constantine the Great, and Theodosius are well-known but not isolated examples of emperors who obviously had a partly conscious grasp of grand strategy, even though some of them may have been misdirected. For example, Trajan's conquest of Dacia and invasion of Parthia were probably too expansive and costly.[10]

Amazingly, considering the ancient Roman interest in warfare, there is very little direct evidence in ancient literary works for Roman grand strategy, but inferential arguments may be made from many different sources and angles. As far as we know, there was no Roman Department of Defense or secretary of the Army of any kind of permanent, institutionalized military council or general staff whatsoever, at least until the late empire.[11] That is particularly surprising since the emperors did have swarms of bureaucratic officials performing many other governmental functions, such as the oversight of fiscal policy. In the days of the Republic the Senate may have served to oversee grand strategy, but under the emperors that was surely not the case.

Despite this lack of direct evidence, I believe it is possible to describe Roman grand strategy and to identify what seem to be the critical considerations in imperial defense. To a certain extent, it will be necessary to repeat arguments in my own book *The Fall of the Roman Empire: The Military Explanation*, but I shall try to keep that to a minimum since I offer some other ideas not set forth or developed there.

First, it should be noted that, contrary to common opinion, the grand strategy of the late Roman Republic was not entirely ad hoc. The Romans in the age of Caesar and Pompey did actually maintain a permanent, standing army, at least in the sense that there were several provinces that regularly required armies in peace or in war.[12] Legions were stationed at various strategic points around the empire. It is true that when Rome became involved in a great war, such as those against Mithridates under Sulla, Lucullus, and Pompey, or the one in Gaul under Caesar, there were special levies, and new armies were created for the occasion. The Roman Senate normally directed the grand strategy of the empire, but the people in assembly could and occasionally did intervene and take matters into their own hands. Commonly this popular intervention was not actually on the level of grand strategy, where the citizens in assembly only rarely interfered with the Senate. More often it was in the selection of a popular leader to serve as commander in chief that the people exercised their authority.

When Augustus gained sole control of the empire in the last half of the first century B.C., he made major changes in Roman grand strategy.[13] Although he may originally have had the rather simple idea that he would merely move out with his armies and conquer the whole world, including Germany, he eventually decided to settle only for that part of the world that was worth conquering. He does not admit even that limitation publicly in the autobiographical account of his reign, the *Res gestae*, where he simply claims to have conquered everything in sight, a process he refers to as "pacification" of the entire world, the *orbis terrarum*.[14]

For the purposes of propaganda the pacification of the world was the new grand strategy, because the Roman concept of pacification clearly entailed subjection to the will of Rome, something that usually happened, in the Roman view, with love, respect, and admiration for the new rulers after initial stubborn resistance had been broken down by the legions. But in the twentieth century we know that grand strategy can be covered with a veneer of pronouncements for public effect that can at times be highly misleading, and there is no reason for us to assume that Augustus always believed his own propaganda. By the end of his reign, when he wrote the *Res gestae*, three legions had been lost in Germany at the Teutoburg Forest, and the emperor, in association with his colleague and adopted son, Tiberius, had already decided to abandon the plan for the conquest of Germany and to make the Rhine Rome's northwestern frontier.[15] When Augustus wrote that the empire was bounded by the ocean from Spain to the mouth of the Elbe, he was not stating a fact—he was concealing a defeat. Nor did he believe that the area north of the Danube had in any effective way been conquered by Roman arms, although he made the claim.[16]

In fact, the line of the northern frontier along the Rhine and the Danube had been reasonably firmly fixed by the death of Rome's first emperor. It was not a particularly flexible line, nor was it defended with much help from client kings. If I may borrow a now famous concept in the analysis of Roman grand strategy, the northern frontier of the early Roman Empire, throughout the first two centuries A.D., did not represent a system of "preclusive security" but a system of "defense in depth." It was not defense in interior depth of the sort applied in the late Roman Empire—it was defense in exterior depth. The Romans considered the Rhine and the Danube as the frontier of the empire, but they defended that frontier by moving across it with their armies whenever it was threatened, using their own frontier as a strong fallback position should it become necessary but relying on their legions to destroy the enemy's main force before it could reach the borders of the empire. Even in the second century Hadrian's Wall was no Maginot line—it was intended as a base for forays forward into hostile territory. True, it was designed to be defended if necessary, but the impenetrable bulwark against the savage Picts was the army, not the wall.[17]

Modern studies of the Roman army have all too often missed this fundamental point. Twentieth-century writers have told us much about what Roman soldiers ate, how they rose through the ranks, what kind of relations they had with women and with their comrades in arms, with whom they slept and in what kind of quarters, what they drank, and how much they were paid. The sociology and the archeology of the Roman army have been the favorite topics of twentieth-century historians of ancient Rome.[18] But little has been written in our century about how they fought, and their manner of fighting is their most distinctive characteristic.[19] The cornerstone of Roman grand strategy, at least in imperial times and particularly along the northern frontier, is the tactical superiority of the legions over all their foes.

In the nineteenth century this point was clearly understood, and it is the basic assumption of one of the most important classics in military history, the great book by Ardant du Picq *Battle Studies, Ancient and Modern*, reprinted recently in volume 2 of *The Roots of Modern Strategy*.[20] Over the centuries, from the earliest days of the primitive city on the Tiber, Romans had been forced by circumstance and guided by the civilizing influences of the Greeks, Etruscans, and Carthaginians to develop military institutions so severe that no modern society could accept them, except in the most extreme circumstances. Romans lived with those institutions constantly, regardless of circumstances in peace or war. Decimation is the most famous example of Roman discipline and training, but it was not in fact often used. The brutality of the centurions was a common complaint of Roman legionaries, however, and what was practiced regularly in a Roman army would cause a public outcry if it occurred today in the U.S. Marine Corps.

Roman discipline and training produced a *fantastic* army. In Du Picq's words, "The determining factor, leaving aside generals of genius, and luck, is the quality of troops, that is, the organization that best assures their spirit, their reliability, their confidence, their unity. . . . We have seen that man will not really fight except under disciplinary pressure. . . . The purpose of discipline is to make men fight in spite of themselves. No army is worthy of the name without discipline. . . . Discipline cannot be secured or created in a day. It is an institution, a tradition."[21]

Du Picq argued that the combination of Roman military discipline with the system of legionary tactical organization made the Roman army the most effective in the history of the world. This rigid system has actually sometimes made the Romans a target of ridicule. In an extreme example, Will Cuppy, in The *Decline and Fall of Practically Everybody*, poked fun at Roman training: "The Romans and Carthaginians were very different in character and temperament. The Carthaginians had no ideals. All they wanted was money and helling around and having a big time. The Romans were stern and dignified, living hard, frugal lives and adhering to the traditional Latin virtues, *gravitas, pietas, simplicitas*, and adultery."[22]

It is important to remember that the Romans undoubtedly exaggerated the severity of their military training, but it was certainly very rigorous. The classic statement in ancient literature of Roman tactical superiority is a speech given by the general and future emperor Titus,

found in the works of the Jewish author Josephus. Titus was address-
ing his troops, and to inspire them he reminded them of their
superiority:

> Now these Jews, though they be very bold and great
> despisers of death, are but a disorderly body, and un-
> skillful in war, and may rather be called a rout than an
> army; while I need say nothing of our skill and our good
> order; for this is the reason why we Romans alone are
> exercised for war in time of peace, that we may not think
> of number for number when we come to fight with our
> enemies; for what advantage should we reap by our con-
> tinual sort of warfare, if we must still be equal in number
> to such as have not been used to war! Now it is not the
> multitude of men, though they be soldiers, that manage
> wars with success, but it is their bravery that does it,
> though they be but a few; for a few are easily set in battle
> array, and can easily assist one another, while over-
> numerous armies are more hurt by themselves than by
> their enemies.[23]

Romans normally fought in close order in waves of thin lines, avoiding the use of the so-called heavy battalions such as the Greek phalanx. The advantage of the Roman tactical system, employing the
famous checkerboard order of battle, was that all available manpower
could be brought into direct action along the battleline. There was no
wastage at the rear of a deep formation. Furthermore, Roman soldiers
were not expected to fight to the death before being replaced by men
from the rear. There was a regular rotation of fighting waves. Obviously, such a system demanded good fighters throughout as each
wave took its place in the front. In heavy battalions, on the other hand,
weak troops can be put in the center of the formation with good troops
in front and rear (panic almost always starts in the rear). There was no
place for weak troops in a Roman legion.

There are many implications for grand strategy in the tactical organization of the Roman legions. In the first place it made possible an
efficient use of manpower, and it gave the Romans tremendous psychological advantages over their foes, but the psychological effect on
Roman soldiers themselves was as important as the effect on the enemy. Because legionaries in the front of the line could expect to be
reinforced in the course of fighting, they fought confidently, and knew
in the depths of their souls that their comrades-in-arms to the rear
would not leave them in the lurch. As a result, contrary to the rule in
premodern military history, Romans inflicted heavy casualties even

when they were defeated. Normally Romans did not run (which is when the heaviest casualties are taken). Against untrained troops, they simply could not be defeated, even when they were greatly outnumbered. Only when a Roman army was caught by surprise on unfavorable terrain, that is, when there was a great failure of generalship, did barbarians have a chance to win a tactical victory. It was this tremendous tactical superiority that made the Roman grand strategy of defending defined frontiers a reasonable use of military manpower.[24]

Grand strategy is an elusive concept, and it normally includes political, diplomatic, economic, and sometimes even religious means of defending the state. The army is not everything. The Roman method of frontier defense in the first and second centuries A.D. did require political stability, since there was no central reserve, and the legions were stationed far from the imperial capital where the emperor normally resided. In the third century, when that political stability broke down, so did the frontiers.

We shall return to that in a moment, but first it might be worth looking at the economic basis of Roman grand strategy. During the Republic, war more than paid for itself as the Romans conquered the richest areas of the ancient Mediterranean, but in imperial times places such as Britain, the Rhineland, the area immediately south of the Danube, and even Dacia to the north did not return to the imperial treasury the cost of conquest or defense.[25] In recent years two authorities on Roman history, working independently, have arrived at very nearly identical estimates of the annual expenditures of the Roman Empire, which would be the same as the annual revenues, for the second century A.D.—somewhere in the neighborhood of eight hundred million sesterces.[26]

About half of the annual expenditures were on the military budget, yet the frontier provinces could not have provided more than 20 percent of the cost of their defense. Naturally, the actual figures varied enormously from province to province. In Imperial times Egypt and Syria, which were frontier provinces, more than paid for the legions stationed in them, but Britain, Lower and Upper Germany, Raetia, Noricum, Pannonia, and Upper and Lower Moesia did not, nor did the gold of Dacia stretch far enough to make a long-term difference in the annual revenues of the Roman Empire.

Because Roman tactical organization and training made it possible for relatively small Roman forces to defeat much larger enemy ones, Roman grand strategy was remarkably cost-efficient.[27] It had to be, because taxes in the early Empire were not particularly high—in the neighborhood of 10 percent of income or produce—and the emperors were extremely reluctant to raise taxes. Confiscation and debasement of coinage were their main means of increasing revenues, and since both measures had drastic political or economic effects, it was important to keep expenditures down.

Since the military budget was naturally larger than any other sector of governmental expenditures, and since the pressures to increase that part of the budget were always great—either to raise legionary pay or to add new legions—the tactical excellence of the Roman army had fiscal as well as military implications. Both considerations were important aspects of Roman grand strategy in the early empire.

Up to this point I have spoken mainly about Rome's northern hinterland along the Rhine and the Danube. In fact, that was the most important frontier by far, and it was the breach of that defensive perimeter that eventually led to the fall of the Roman Empire in the west. The frontier to the south, the great desert of the Sahara, needed little defense and was never seriously harassed, though some police action was occasionally necessary there.[28] Roman North Africa could be threatened, if we set aside the possibility of rebellion, only by invaders across the northern or eastern frontiers. When Roman Carthage finally was conquered in A.D. 439, it fell to Vandals who had crossed the Rhine only a generation earlier. So in the end it was the north that mattered, and more than half of all Rome's legions were stationed along that great line from Britain to the Black Sea.

The *eastern* frontier required a grand strategy that was different than in the north in that diplomacy played a greater role, and was, sometimes at least, as important as military policy.[29] In the east in the first two centuries A.D. the great threat came from the Parthians, a civilized people, heirs to the Seleucids and to the earlier Persians. The Parthians had much better armies than the northern barbarians, and they also had an organized government with a reasonably sophisticated bureaucracy. They had what we would call ambassadors, though not permanent ones, but they knew how to conduct formal negotiations and understood the importance of keeping their word. Equally important, Parthians were not migratory. The line between Parthia and Rome was clear. Of course, there could be war between the two great empires, and encroachment from either side on the frontiers, but Parthians were not people who simply wanted to move their entire population—men, women, and children—into new territory. As a result, conditions in the east were fundamentally different from those of the north, and in some ways easier for the Romans to control.

There were usually four legions stationed in Syria, a force altogether, including auxiliaries, of nearly forty thousand men. These troops had a twofold purpose: to guard the eastern frontier against Parthia and to deal with disturbances in Judaea if necessary. Roman emperors carefully stationed this rather large army outside of Palestine, yet very nearby. Whenever the prefect in Judaea needed military assistance in addition to the small forces under his immediate personal command, he could count on support from the governor of Syria.[30] (Luttwak's argument that the Roman siege of Masada "reveals the exceedingly subtle workings of a long-range security policy," or "a calculated act of psychological warfare," misses one of the most important points of Roman grand strategy and gives Romans credit for more sophistication and subtlety than they had. The Roman commander who ordered construction of the ramp up to the top of the strongly defended mountain fortress was merely doing his job in the simplest and most direct fashion he could, relying on another of the great strengths of the legions, military engineering, combined with the discipline and training of the troops.)[31]

In fact, in all the support branches of the armed forces the Romans had a tremendous advantage over their enemies, at least in imperial times. Their logistical system was outstanding, from the stable-fed horses of the late Roman cavalry to the governmental armories to the food supply system for the men in the field. In the fourth century Constantius II, son of Constantine the Great, faced rebellion by his nephew, later known as Julian the Apostate. Julian had been based in Gaul, and Constantius prepared to march west from the eastern empire to face him. The emperor's agents gathered three million bushels of wheat along the borders of Gaul and another three million further east so that there would be depots of ample supplies as Constantius led his troops into battle. A famous historian of late antiquity and the early Middle Ages, E. A. Thompson, has observed, "When an army of northern barbarians undertook a campaign, its leaders did not think in terms of millions of bushels of wheat."[32]

The Romans had so many similar advantages over the barbarians to the north that it would be futile to discuss them all. One must be mentioned, however: the Roman edge in siege warfare. Except for the Huns, barbarian armies could not conduct effective sieges. Fritigern, one of the earliest known leaders of the Visigoths, said that the Goths kept peace with walls, a euphemistic maxim that meant that Goths could not take fortified cities.[33] When Rome fell to Alaric in A.D. 410, it was because someone on the inside opened the gates.[34]

Before I turn to the grand strategy of the late Roman Empire, one or two points about the earlier period need emphasis. Rome's defensive policy in the first two centuries A.D. demanded dynastic political stability, as we have seen, but it also relied upon the support and loyalty of the interior provinces. Except for Spain briefly, and Egypt and Judaea more or less permanently, the emperors were free to deploy the legions for frontier defense. The fact that people of disparate cultures and languages all around the Mediterranean became essentially loyal Romans was an important condition in enabling the emperors to concentrate their attention on the Rhine, the Danube, and the Parthians.[35]

This imperial loyalty also made possible the extensive use of auxiliaries alongside Roman legions. Only citizens could serve in the legions, but non-Roman provincials were recruited for the auxiliary units, which in total strength equaled and may actually have exceeded the number of citizens under arms. Without the support of these auxiliaries Roman strength would have been reduced in many significant ways.[36] Most of the cavalry and the numerous skirmisher units that supported the legions in the field were provided by auxiliaries. To that extent the propaganda that encouraged loyalty to the imperial ideal as well as the generally good government that helped to produce it can also be considered features of Roman grand strategy. Roman emperors, at least the good ones, and many that were not so good, understood this point.[37]

Loyalty to the imperial ideal was important in another, though related, way. Throughout the history of the empire it was difficult to find an adequate number of volunteers for military service, and compulsion was frequently required to make up for this shortage.[38] In the late empire we know that the height requirement for service in the army was reduced. To avoid service, prospective draftees cut off their own thumbs, and the emperors responded at one point by making that a capital crime.[39]

But the manpower problem was actually as great in the early empire. Most people know the famous story from the biographer Suetonius that Emperor Augustus, when he learned that his general Varus had lost three legions in Germany, banged his head against the wall and cried out, "Varus, Varus, give me back my legions." Why did he react that way? The reason is that draft resistance was so great in the last years of his reign that he could not find replacements for the lost troops. Neither Romans nor provincials were interested in any further wars of conquest along the Rhine and the Danube. Rome had maintained twenty-eight legions before the defeat in the Teutoburg Forest, but afterwards, for many years until the reign of Caligula, or perhaps Claudius, the loss was not made up, and the number of legions remained at twenty-five.[40]

It was not, as some have assumed, a matter of money—the imperial budget would have permitted replacing those three legions,—but Roman citizens could not be found to serve in them. Augustus actually enrolled former slaves in the legions in the last years of his life.[41] This is amazing, since we know from the imperial census taken in A.D. 14 that there were nearly 5,000,000 citizens altogether. In twenty-five legions there would have been only about 125,000 citizens under arms, and an additional three legions would have required merely 15,000 more, plus an equal number of auxiliaries. As a percentage of the total number of citizens this figure is minuscule, and it reflects how serious the problem of conscription was.

As we have noted, the third century witnessed a terrible collapse of central authority, the frontiers were overwhelmed, and the empire almost collapsed. One emperor in this period, Valerian, was actually captured by the Persians, who under a new Sassanian dynasty represented a far greater threat to the eastern frontier than their predecessors, the Parthians. Another emperor, Decius, died in battle fighting the Goths on the Danube.

With the breakdown of the frontiers and the splintering of the empire in the hands of various rebels, traditional Roman grand strategy no longer prevailed. In the sense that there was no more a unified empire, there was no imperial grand strategy. It is sometimes said that Gallienus, in the middle of the century, organized a mobile central army based on a strong cavalry, and operated out of Milan in order to be closer to the troubled frontiers, but whatever Gallienus did—and there is not much evidence—his policies were not lasting and had evaporated by the end of the century and the reign of Diocletian.[42] To most of the emperors of the mid-third century the old frontiers were remote, and it was impossible to project any significant military power to them. When Diocletian restored the military integrity of the historic frontiers, he did so with essentially the same legionary army that had served Rome in the days of the high empire. His grand strategy was the old one of preclusive security or the defense of a well-defined frontier. Luttwak, following Theodor Mommsen, believes that Diocletian may have originated the later system of defense-in-depth, but he calls it "shallow" defense in depth, though shallow depth might mean almost no depth at all.[43]

In fact, Diocletian did not change traditional Roman grand strategy— he reintroduced it. The responsibility for the change rests directly with Constantine the Great. In what may be the most straightforward statement in ancient literature on grand strategy, the fifth-century historian Zosimus said:

> Constantine abolished this [frontier] security by re-
> moving the greater part of the soldiery from the frontiers
> to cities that needed no auxiliary forces. He thus de-
> prived of help the people who were harassed by the bar-
> barians and burdened tranquil cities with the pest of the
> military, so that several straightway were deserted. More-
> over, he softened the soldiers, who treated themselves to
> shows and luxuries. Indeed (to speak plainly) he person-
> ally planted the first seeds of our present devastated state
> of affairs.[44]

The creation of this central reserve deep in the interior of the empire proved to be the biggest change in the history of imperial grand
strategy. The large mobile field army of a hundred thousand or more
was created by withdrawing units from the frontiers, and Edward
Gibbon, following Zosimus, excoriated the first Christian emperor for
corrupting military discipline and preparing the ruin of the empire.[45]
But Mommsen and Luttwak, and almost everyone else, have argued
that the new system of defense was more realistic than the earlier one
and theoretically sounder because it relied on a central reserve.[46]

Defense in depth, however, has many flaws. It almost automatically
permits penetration of the frontiers by invaders, and can protect them
only by the threat of massive retaliation by the central, mobile army.
Indeed, under Constantine and his sons the frontiers remained secure, but in the long run they did deteriorate. A rapid deployment
force today can actually be moved many places in the world quite
quickly, but in antiquity even a mobile army needed about two
months to move from Rome to Cologne.[47] Furthermore, the emperor
might have been comforted by the presence of a field army near the
court, but Roman subjects on the frontier did not appreciate the advantage. The reduction of military strength where it appeared to be
most needed seemed more a threat than a benefit to inhabitants of the
frontier provinces.

Obviously, the worst feature of defense in depth is that the central
mobile army becomes an elite force while the frontier defenders serve
a secondary role in defense policy.[48] Troops that are not expected to
defeat the enemy—merely to delay and harass him—cannot be
blamed for falling back quickly to let the crack troops move up with
the killing blow. Over the course of a century after Constantine's
change in Roman grand strategy, the frontier troops turned into a
ragtag local militia while the mobile armies did all the major fighting.
Since the border guards were so ineffective, Rome's military man-

power problems were greatly intensified. The army of the late empire was actually larger than that of the second century, but only part of it was tactically competent. Roman infantry, which had been the bulwark of military defense since the beginning of Roman history, was sadly undermined as cavalry emerged to a favored position in the mobile army of the late empire.

The best source for the decline of Roman infantry is the famous treatise of Vegetius, *De re militari*, or *On Military Matters*, written in the western empire in the fifth century. In the Middle Ages and early modern times, it was one of the most popular works of ancient literature. Approximately 150 manuscript copies have survived, and even before the invention of printing it had been translated into English, French, and Bulgarian. In the earliest days of printing, between 1473 and 1489, it was published in five different countries, but today interest in it has declined, and the only English translation that is readily available is in volume 1 of *The Roots of Modern Strategy*, though it is not complete.[49]

It is a manual that describes the earlier Roman system of conscription, training, strategy, and tactics, a system that had been so entirely forgotten by the fifth century that, according to Vegetius, no one knew much about it and everything had to be learned from books. It is fashionable today to dismiss Vegetius as unsophisticated, partly on the grounds of style, although one of Napoleon's marshals once said, "I don't know why his Latin is not liked; I myself like it very much, because I understand it."[50] In fact, Vegetius is clear, stylistically and substantively. He saw Rome's greatest problem in the decline of infantry. In one passage he says that he is not even going to discuss cavalry, because in cavalry Rome was competitive with her enemies. Infantry was the difficulty.

According to Vegetius, down to the death of Emperor Gratian in 383,

> footsoldiers wore breastplates and helmets. But when, because of negligence and laziness, parade ground drills were abandoned, the customary armor began to seem heavy, since the soldiers rarely wore it. Therefore, they asked the emperor to set aside first the breastplates and mail, and then the helmets. So our soldiers fought the Goths without any protection for chest and head and were often beaten by archers. Although there were many disasters, which led to the loss of great cities, no one tried to restore breastplates and helmets to the infantry.

> Thus it happens that troops in battle, exposed to wounds because they have no armor, think about running and not about fighting.[51]

What caused this decline in Roman infantry—once the proudest footsoldiers in the world? Partly it was the effect of barbarism on the army. Constantine was the first emperor to use Germanic troops in Roman service on a large scale, although some barbarian mercenaries had been used regularly since the Republic. But by the end of the fourth century Theodosius fought the battle of the Frigid River with twenty thousand Visigoths in the vanguard of his army, and from that time forward barbarians formed a significant percentage of troops in Roman service, even in the central mobile armies. Sometimes more than half of a Roman field army was made up of barbarians.

Usually these Germanic troops served under their own tribal commanders and were subject to their own discipline, which was lax. Worse than that, they were probably better paid for their services than Roman citizens, who readily resented the differences between their own harsh discipline and low pay and the relatively comfortable conditions under which the barbarians served. To avoid mutiny, Roman commanders had to agree to soften the training of the infantry, particularly the frontier guards, who had only a secondary role in the unfolding of Rome's grand strategy in any event. In the end it was the combination of the barbarism of the army and the adoption of a fundamentally flawed grand strategy that led to the loss of the western Roman Empire in the fifth century. The decline in the quality of what would later be called the "PBI," the poor bloody infantry, and the lack of definable frontiers, both results of the new strategy of defense in depth, made it impossible to save the Roman Empire.[52]

Managing Decline: Olivares and the Grand Strategy of Imperial Spain

J. H. Elliott

Writing in the *New Republic* in 1985, in an issue devoted to the problems of contemporary British society,[1] Peter Jenkins quoted from my book *Imperial Spain*, published twenty-eight years ago:[2] "Heirs to a society which had over-invested in empire, and surrounded by the increasingly shabby remnants of a dwindling inheritance, they could not bring themselves to surrender their memories and alter the antique pattern of their lives. At a time when the face of Europe was altering more rapidly than ever before, the country that had once been its leading power proved to be lacking the essential ingredient for survival—the willingness to change." I was referring, of course, not to the British ruling class of the twentieth century but to the Spanish of the seventeenth. The use of the Spanish case in this context, however, by a British journalist suggests that it may have a certain contemporary relevance, and that there are perhaps recurring situations in the trajectories of great powers which raise central issues of policy and practice that cut across the centuries.

My point in this quotation was that the ruling class of seventeenth-century Spain was too hidebound, too traditional in its attitudes and values, to adjust to the realities of declining power. But this raises a question that goes beyond the history of Spain—a question implied, indeed, in those loaded words (not of my own choosing) *managing decline.* How can, and should, a great power "manage decline"? Indeed, can decline be "managed"at all, and if it can, what strategy for management is likely to work best? Or is decline a phenomenon of nature that defies man-made management, and is the "management of decline" no more than a sophisticated euphemism for making the best of a bad job?

Any general theory of decline which is to have some hope of commanding widespread assent will need to rest on a careful analysis of individual case histories.[3] One such case history, which is now relatively well documented, is that of Spain in the 1620s and 1630s, during the ministry of Don Gaspar de Guzmán, who has gone down to history under the name of the count-duke of Olivares.[4] The sixteen-year-old Philip IV of Spain, who entrusted the fortunes of Spain to Olivares on his accession to the throne in 1621, succeeded to an impressive but troubled inheritance. The rise of Spain as a great European power is conventionally, and conveniently, dated to the union of Castile and Aragon following the marriage in 1469 of Ferdinand of Aragon and Isabella of Castile. With the reconquest of the Moorish kingdom of Granada and Columbus's landfall in America in 1492, the way was opened for Spain's great imperial expansion of the sixteenth century. When in 1556 Philip II succeeded his father, Emperor Charles V, as king of Spain, he was the ruler of dominions which stretched from the Netherlands and Spanish Italy (Naples, Sicily, and Milan) to the Pacific coast of America and its great viceroyalties of Mexico and Peru; and to this vast agglomeration of territories he added, in 1580, Portugal and its overseas empire in Brazil and the Far East, as the result of his succession to the Portuguese throne.

In spite of its impressive military and administrative achievements, the Spanish empire (or *monarchy,* as it was known to contemporaries) was showing obvious signs of strain during the second half of the reign of Philip II. There were growing indications in the 1580s and 1590s that the monarchy was over-stretched, or in other words that its resources, in terms of money and manpower, were proving unequal to the tasks that it was being called upon to face. What were these tasks? They have to be seen in the general context of the goals of the dynasty: in the first instance, the maintenance of order and justice in the king's numerous dominions, and their protection from enemy attack; second, the upholding of Catholic Christianity against Islam and against the rising tide of Protestant heresy and subversion in Europe; and third, the maintenance of the closest possible unity between Madrid and Vienna as the best means of guaranteeing the dynastic interests of the Spanish and Austrian branches of the house of Habsburg and the stability and survival of the traditional ordering of Christendom.

Within this overall context of its perceived goals and aims, Spain was confronted during the reign of Philip II by a series of challenges, of which the most acute was the revolt of the Netherlands in the 1560s. The failure to suppress this revolt at an early stage committed Spain to a prolonged military struggle in the Netherlands which imposed enormous strains on its resources. That great military machine, Spain's army of Flanders, built up in the loyal provinces of the southern Netherlands to contain and, if possible, suppress the rebellious Dutch, came to have an establishment of some fifty to sixty thousand men, and its expenses were to be of the order of three million ducats a year—approximately a third of the crown's total annual expenditure.[5] But even such a heavy commitment of men and money failed to crush the revolt. Instead, the war against the Dutch escalated in the 1580s and 1590s into a great northern struggle on land and sea between Spain and the forces of international Protestantism—the Dutch, the Huguenots, and the England of Elizabeth. By the time Philip II died in 1598, Spain was visibly faltering in this northern struggle. Philip himself, with the crown temporarily bankrupt, sounded the beginnings of retreat in the north when he made peace with France in 1598; and the process continued under his successor, Philip III, who reached a peace agreement with the England of James I in 1604, and reluctantly, in 1609, signed a Twelve Years' Truce with the rebellious Dutch. The Dutch truce was followed by a decade of relative, and uneasy, peace in Europe—the age of the Pax Hispanica, when the sudden weakening of French power following the uncovenanted bonanza of the assassination of Henry IV in 1610 left Spain, at least temporarily, in a position of undisputed European primacy.[6]

This change at the turn of the century from bellicosity to relative pacificism—from a generation of war to a generation of peace—was really dictated by force majeure, and did not imply any abandonment of Spain's long-term goals and aspirations. The force majeure was essentially financial and economic. Expenditure in the last years of Philip II had been dramatically outrunning resources—resources which depended essentially on the willingness and ability of the international banking community (especially the Genoese) to continue making loans to the Spanish crown; on the capacity of the silver mines of Mexico and Peru to go on producing a regular supply of bullion for the king, to the tune of two to three million ducats a year; and on the tax-paying capabilities of the king's various dominions and provinces, and especially Castile, which under Philip II had become the center and heart of the monarchy, and which bore the brunt of the financial and military burden imposed by the foreign policy requirements of the dynasty.

But from the 1590s the alarm bells were ringing in Castile. There were growing signs of agrarian and demographic crisis, especially after the great plague at the end of the century, and an increasing sense that the weight and distribution of taxes were systematically crippling the most productive sectors of Castilian society. The reign of Philip III, from 1598 to 1621, saw an extraordinary outpouring in Castile of tracts and treatises which attempted to diagnose the ills from which Castile was suffering and to prescribe appropriate remedies. This economic and political literature, produced by men who came to be known collectively as *arbitristas* because they proposed *arbitrios*, or expedients, for the salvation of the realm, took as its starting point the sickness or decline of the body politic. In fact, the word *declinación* seems to have been used for the first time in 1600 in an extremely influential treatise by one of the first and most intelligent of the arbitristas, Martín González de Cellorigo, and the concept of decline was to take hold of the Castilian consciousness in subsequent years.[7]

This contemporary casting of the problems of Castile and the Spanish monarchy in terms of decline seems to me of enormous significance if we are to understand the strategy that would be adopted by Olivares and the government of Philip IV in the 1620s. "Decline" had two connotations: biological and historical. Biologically, every organism went through a process of growth, maturity, and decay; and since the process was natural, it was irreversible—unless, of course (and this was the one saving possibility) God should intervene directly in the affairs of men. This situating of Spain's past, present, and future within the cyclical process of the natural world seems to me to have played its part in creating that sense of fatalism which is so perceptible in the Spanish ruling class of the seventeenth century. The wheel was inexorably turning, and the best one could hope to do was to hold it back for just a little longer before its next downward turn. Pessimism was reinforced by the historical analogy of the fate of the Roman Empire. Spaniards in the days of their greatness had seen themselves as the modern heirs of the Romans; and now, in the days of trouble, the analogy they had so confidently borrowed became distinctly uncomfortable. If they were indeed latter-day Romans, how could imperial Spain hope to escape the process of degeneration and decline to which imperial Rome had finally succumbed?

But no true believer could fully accept the determinist implications of the concept of decline, and room therefore had always to be found for the possibility of divine intervention. This was where the story acquired a *moral* dimension. Seventeenth-century Spaniards who drew the parallels between Rome and Spain were well aware that the

Roman Empire had been subverted from within by the decline of moral standards—by luxury, greed, effeminacy, and the abandonment of the traditional moral and martial virtues that had given Rome its greatness—and they saw a similar process at work in their own society. This meant that, alongside their elaborate schemes for economic and fiscal reform, they also campaigned for a reform of manners and morals as being at least equally important for the salvation of their country. Only the abandonment of public and private vices, and moral renewal, could "oblige" God, in the seventeenth-century phraseology, to come to the rescue of His chosen people and save them from the otherwise irreversible degenerative process to which, like all natural and political bodies, Spain was prone.

In the first two decades of the seventeenth century, therefore, we see the creation of an intellectual or mental climate in which the two predominant elements of *pessimism* and *reformism* were more or less equally balanced. Spain was sinking—but it might yet, if it followed the right course, save itself. But everything that happened during the reign of Philip III suggested that the right course was very far from being followed. The government of the favorite of Philip III, the duke of Lerma, was corrupt and incompetent; the years of peace, which might have been used for fiscal, administrative, and social reform, were being scandalously wasted; and foreign policy problems, which had temporarily gone into abeyance, were again coming to the fore. In particular, the Bohemian revolt of 1618 threw into jeopardy the whole position of the Catholic cause and the house of Austria throughout Central Europe; and the Twelve Years' Truce with the Dutch was drawing to its close, raising the whole issue of whether to renew, or go back to war. At the same time, Spain's capacity for war, whether in Central Europe or the Netherlands, was being reduced still further by a slump in the revenues from the Indies. In 1620 the crown received only eight hundred thousand ducats from America, against the two million ducats a year it had been receiving at the beginning of the reign.

The combination of domestic crisis with foreign policy challenges came to a head in 1618–1621, the last three years of the reign of Philip III. On the one hand, there was a growing movement at home for drastic reform—a movement supported by the Cortes of Castile, by sections of the administration and the urban patriciates, and by a public opinion informed by the writings of the arbitristas and alarmed by the growing indications of national decline. On the other hand, there was mounting exasperation among Spain's diplomats, generals, and foreign proconsuls at the feebleness of Lerma's foreign policy and the humiliations to which it was subjecting the Spanish crown. These men, like Don Baltasar de Zúñiga, who returned to Spain from the embassy in Prague in 1617 and took up a seat on the Council of State, were activists and interventionists who wanted to restore Spain's waning reputation before it was too late. It was this combination of domestic and foreign policy activists which toppled the duke of Lerma in 1618 and finally secured uncontested power in 1621 in the persons of Don Baltasar de Zúñiga and his nephew, the count-duke of Olivares.

Assumptions and Attitudes

It would therefore seem fair to say that the "grand strategy" of the Olivares years, from 1621 until his fall from power in 1643, was to a large extent set by the conjunction of forces, and the attitudes of mind, that emerged during the preceding two decades. In that sense, it was a fairly predictable strategy, although what was not predictable was the extraordinary energy, stamina, and resourcefulness that Olivares himself would display in his deployment of this strategy. The elements of the Olivares program are all there, at least in embryo, before Olivares himself comes to power, even if the final combination of elements was clearly shaped by the priorities of Olivares and Don Baltasar de Zúñiga. Of what did their grand strategy consist?

In the first place, I do not think we can fairly describe it as a grand strategy for the management of decline. The fact of decline—administrative, military, economic, fiscal, demographic—was admitted and was the starting point for action. Olivares himself told the king in 1624: "The present state of these kingdoms is, for our sins, quite possibly the worst that has ever been known."[8] But the process of decline had to be checked, and, if possible, reversed. In other words, the overwhelming need of the moment, again to use his own words, was to "resuscitate Your Majesty's monarchy."[9] His program, therefore, was conceived as a program for the regeneration and revival of the Spanish monarchy and empire, beginning with the regeneration and revival of its heartland, Castile.

From a twentieth-century standpoint, knowing what we know of the fate of imperial Spain, our diagnosis would surely be that this was a society suffering from the cumulative effects of a long-standing failure to adjust its foreign policy aims to shrinking imperial and national resources; and our prescription would presumably consist of some combination of the following elements: a prolonged period of peace, a systematic attempt to promote economic revival, and a reassessment of priorities as part of a long-term effort to tailor ends to means. But this, which seems the rational approach to us from our twentieth-century vantage point, reflects a scale of values very different from those of the seventeenth century; and it also, I believe ignores the untidy realities that in any period are associated with the possession of worldwide empire.

As regards differences in the scale of values between the seventeenth century and our own, one of the most critical is surely the fact that the avoidance of war, which many would now see as the supreme aim of the great world powers, was then much lower in their scale of priorities. War was accepted as a natural fact of life, and the object of grand strategy was simply to go into war at such a time and in such a way as to ensure the maximum chance of victory. The seventeenth-century world was a world in which "reputation"—or, in modern parlance, prestige, or the saving of face—was paramount, and reputation could only be sustained by the victorious display of military power. This means that in the discussions of the Spanish Council of State between 1619 and 1621, which culminated in the fateful decision to let the truce with the Dutch expire and resume hostilities in the Netherlands in spite of Spain's acknowledged financial and economic weakness, we should not be surprised by the kind of arguments that were deployed. It was argued, with some justice, that peace with the Dutch had proved even more disastrous than war: that the savings on the army of Flanders had not been so great as to compensate for the heavy blows inflicted by the Dutch on the overseas possessions of Spain and Portugal under the cover of a nominal peace. But it was also argued, not least by Zúñiga, that reputation was the critical element. "In my view," he said, "a monarchy that has lost its *reputación*, even if it has lost no territory, is a sky without light, a sun without rays, a body without a soul."[10] One might, perhaps, dismiss this as a mere rhetorial flight. But Zúñiga was a man of immense experience in international affairs, and a "realist," in the sense that he had no illusions about the possibilities of Spain at this late stage crushing the Dutch revolt. The most that could be hoped for was an honorable peace—a peace less damaging to the king of Spain's reputation than the truce of 1609. Like others of his generation who looked back to the great age of Philip II and had suffered through the humiliations of Lerma's foreign policy, he appreciated that "reputation" was a vital element in the armory of a great power, and that, once its reputation began to slip, all the smaller powers would begin to tweak its nose.

Given this attitude of mind, and the bitterness of the reaction against the pacific policies of the Lerma regime, Spain's return to war against the Dutch in 1621 was virtually preordained. Indeed, some even saw it as a moral necessity. In one of the meetings of the Council of State, the count of Benavente argued that the only result of peace in the Netherlands had been to debilitate Spanish manhood. A martial people was growing effeminate, and, as he said, "either we have a good war, or we lose everything."[11] In fact, we have here a conscious appeal to the traditional martial values which had given Castile its greatness; and this return to the past—going back beyond the immediate and humiliating past represented by the reign of Philip III to the more remote and glorious past of a reign of Philip II that was beginning to be idealized with the passage of time—was to be one of the guiding principles of the Olivares regime. Renewal, in other words, could only come with the restoration of old values. As a result, one major aspect of the Olivares program was to be a great spiritual and moral house-cleansing, a puritanical reform movement which involved the closing of brothels, a ban on the publication of frivolous literature, and a new drive for austerity in dress.

The starting point of the Olivares years, then, was a renewed commitment to war—war that would at once purify and reinvigorate Castile and restore the king of Spain to his proper, and God-assigned, position as the greatest monarch in the world. This position, as Olivares himself recognized, implied either an almost constant state of warfare, or at least a readiness for war. He once compared Spain to a lion in the forest, who is feared and respected, but not loved. The reassertion of Spanish power, which would ensure it due fear and respect, was therefore the central feature in Olivares's program. But how was this to be achieved? The goal was traditional, but the methods must be new. In 1619 Spain's ambassador in London, the count of Gondomar, had commented in a letter to the king on changes in the nature of modern warfare: "Warfare today is not a question of brute strength, as if men were bulls, nor even a question of battles, but rather of winning or losing friends and trade, and this is the question to which all good governments should address themselves."[12]

Olivares was well aware of these changing requirements, which had been brought home to the states of early seventeenth-century Europe by the successes of the Dutch. The Dutch, starting from small beginnings, had shown how power and prosperity were intimately related, and how the intelligent mobilization and development of apparently limited resources could lead to victory in war. The Spaniards, therefore, must take a leaf out of their enemy's book. Just as Olivares turned, then, to the Spanish past for the moral and martial revival of Castile, so he turned to the Dutch present as a model for economic revival and technological advance. War, following the

Dutch example, must move from land to sea. Trading companies must
be founded, industry and entrepreneurship be encouraged, and rivers
be made navigable, so that productivity could be increased. Castile's
cumbersome tax system must be radically overhauled, with the
eventual goal a single consolidated tax, and the constitutional structure of the Spanish monarchy be redesigned to distribute the fiscal
and military burdens more equitably among its component parts. To
use current terminology, Olivares's program, in this aspect at least,
was to be a program of *modernization*, designed to make Spain once
again economically competitive and, in consequence, more effective
in war.

Olivares seems to have at least partially recognized that such a
program required a transformation of social attitudes and values.
Those who at present found themselves marginalized by the taint of
Jewish blood, which excluded them from office and honor, must
somehow be reincorporated into national life, so that the country was
not deprived of their talents. Educational reform was to produce a
new generation of nobles, better disciplined and better equipped than
their fathers for the royal service. To create a society which generated
more wealth, it was necessary, in Olivares's words, to "bend all our
efforts to turning Spaniards into merchants."[13] One could argue that
there was a fundamental contradiction between the attempt to mod-
ernize Castilian society on the one hand, and to revive its traditional
values on the other; but I suspect that this is a contradiction to which
we have grown accustomed in the twentieth-century world. It is possible that the most effective modernizers are those who can cloak their
programs in the rhetoric of the revival of past glories. Whether
Olivares could perform this trick and move Castilian society into a
more modern age without prompting a fundamentalist backlash remained to be seen.

The regeneration of Castile, as Olivares saw it, was dependent not
only on its economic and moral renewal, but also on the development
of a more rational and economical system for sustaining the military
and naval capability of the Spanish monarchy in the new age of total
war. This meant schemes for fixed annual appropriations of resources
for Spain's armies and its fleets; and in 1625 Olivares launched his
great project for a radical overhaul of the whole system of imperial
defense—a project that he called the "Union of Arms." The Union of
Arms was intended to take some of the strain off Castile by allocating
to each kingdom and province of the monarchy—Castile, the crown of
Aragon, Portugal, Naples, Sicily, Flanders, and the Indies—responsibility for the provision and maintenance over a period of fifteen years

of a quota of paid men, the quota being fixed in accordance with their presumed populations. This would yield a potential defense force of 140,000 men, of whom one-seventh would be called to arms any time some part of the monarchy and empire was subject to attack. Given the wide constitutional disparities between the king's various dominions, this was an enormously ambitious scheme, and Olivares ran into trouble almost immediately in his attempts to secure its adoption. But in his eyes it represented the best, and indeed the only, hope for the salvation of the monarchy, and the attempt to introduce it by one means or another remained a consistent and principal element in his policies throughout his years of power.

Policies and Problems

Faced, then, with the challenge of decline, the response of Olivares and his colleagues on the Council of State was not to reassess and scale down Spain's traditional foreign policy objectives, but to reform and rationalize existing structures in the hope of making those objectives more attainable. Indeed, in any clash between the claims of foreign policy and the availability of resources, the predisposition was to assume that the international emergency was too great to be shirked, and that somehow or other the resources would be found. When in 1625, for instance, the Council of State was debating whether it was possible to go on subsidizing the emperor while simultaneously fighting the war in the Netherlands, the marquis of Montesclaros, president of the Council of Finance, remarked that "the lack of money is serious, but it is more important to preserve reputation."[14] This is not the sentiment one usually expects of a chancellor of the Exchequer, and certainly not the one expressed by Harold Macmillan at the height of the Suez crisis in 1956.

Are we then to assume that we have here a last imperial generation, so attuned to past glories as to be totally out of touch with modern-day realities? I have already suggested that we have to beware of imposing our own scale of values on seventeenth-century minds, and that the whole concept of *reputation* had a resonance in the seventeenth century that was not confined to Spain. Cardinal Richelieu, for instance, whom modern historiography is all too inclined to depict as an archrealist in comparison with a quintessentially quixotic Olivares, was just as obsessed as his rival with the need to maintain reputation. For Richelieu as for Olivares, if a choice had to be made between *réputation* and *repos*, it was réputation that won.[15]

To Europeans in the age of Olivares, a Spain in pursuit of reputation was a Spain once again on the march. In the 1620s and 1630s they saw before them the awful specter of a Habsburg-dominated Europe, and a revival under Olivares's leadership of the grandiose designs for world empire that they associated with the reign of Philip II. But what to them was a newly aggressive imperialism took on a very different coloring when viewed from the capital of the imperial power, Madrid. Although the Council of State in the first years of Philip IV still had its share of triumphalists, it also had its fatalists, including Don Baltasar de Zúñiga, who set the course for his nephew's foreign policy. Olivares, in the tradition of his uncle, was, I believe, at heart a fatalist, although perhaps a fatalist with triumphalist moments. To him, as to Zúñiga, Spain's monarchy and empire were under attack from all sides, and unless God should miraculously intervene on Spain's behalf, its long-term prospects were grim. When Gondomar, in a letter of 1625, asserted that the ship was going down, Olivares did not essentially disagree, although he did not want it said out loud for fear of discouraging those around him. "I know it and lament it," he replied, "without letting it weaken my determination or diminish my concern; for the extent of my obligation is such as to make me resolve to die clinging to my oar till not a splinter is left."[16]

What I detect in the count-duke and the more responsible of his colleagues is not so much the arrogance of empire as an almost overwhelming sense of its burdens. The weight of their imperial inheritance lay heavy upon them; they had received the precious charge from their forefathers, and it was their task, if possible, to pass it on in turn to posterity. *Conservación*—conservation—was therefore a critical word in their vocabulary. It was not a question of expanding empire, or gaining more territory, but of preserving to the best of one's ability the territories it possessed and the reputation of its king. In other words, although to most Europeans the policies of Madrid appeared intolerably aggressive, Olivares and his colleagues saw themselves as perennially on the defensive. This defensive mentality was underlined by the belief in an explicitly asserted domino theory. In 1635, on the eve of the outbreak of war with France, he wrote: "The first and most fundamental dangers threaten Milan, Flanders and Germany. Any blow against these would be fatal to this monarchy; and if any one of them were to go, the rest of the monarchy would follow, for Germany would be followed by Italy and Flanders, Flanders by the Indies, and Milan by Naples and Sicily."[17]

The essence of the count-duke's foreign policy was therefore to conserve with reputation a worldwide monarchy that was already large enough. Somehow the line must be held and the dominoes be prevented from falling. His aim, he told his colleagues on the Council of State at the end of 1626, was "universal peace with honor."[18] How was this to be achieved? On the one hand, as I have suggested, by a policy of reforms which would strengthen Spain's military and naval capacity and restore its standing in the world. On the other, by a series of well-defined foreign policy objectives, to be achieved from a posture of strength. France—always a potential threat to Spanish power—must be contained. The Habsburg position in Germany must be restored by giving consistent support to the emperor, but at the same time taking care not to drive the Protestant princes to despair, and so escalating the conflict. Finally, the Dutch—the most critical problem of all—must be brought to the negotiating table by a rationalized system of warfare (including the deployment of Spanish naval power and economic blockade) and persuaded to accept a peace more favorable to Spain than that of 1609.

In practice, it proved much easier to state the objectives than to achieve them. French power under Richelieu was becoming more assertive. The emperor turned out to be much less amenable to control by Madrid than Olivares had expected, and adopted a series of policies which led to the very extension of the war in Germany that the count-duke had hoped to prevent. And the Dutch, although showing signs of being prepared to negotiate under pressure, never quite acceded to terms that he considered sufficiently satisfactory for the conclusion of a settlement. There were moments in the 1620s when his aim of universal peace with honor seemed almost within his grasp—in 1625, for instance, that great year of Spanish victories at Breda, at Cadiz, and in Brazil. These victories, and especially the victory in Brazil, achieved by an expeditionary force of fifty-two ships and twelve thousand men,[19] were a remarkable testimonial to the continuing resilience of Spanish power and to the organizing capacity of the Spanish administrative machine. Yet each time the final prize seemed somehow to elude Olivares's grasp, and at the end of 1627 he made what looks like his cardinal mistake in the field of foreign policy by deciding that Spain should intervene in the Mantuan succession.

The count-duke's decision to intervene in Mantua is perfectly comprehensible in the context of the times. Some strong arguments could be advanced, simply in terms of the Italian and international situation, against allowing the duke of Nevers, the candidate with the strongest claims, to take over the inheritance unopposed. The fact that the strategically placed duchy of Mantua would now be in the hands of a French-born duke presented obvious risks to Milan and to Spain's general position in northern Italy. It could also be argued that, by precipitately claiming his inheritance without reference to his over-lord, the emperor, Nevers had struck a damaging blow at imperial authority. Yet these and other questions could probably have been resolved by diplomacy if Madrid had so wished. It is hard to believe that Olivares's decision in favor of military intervention was unrelated to his current difficulties at home, where his reform pro-gram was being blocked by strong domestic opposition. A spectacular success, and notably the permanent occupation by Spain of the for-tress of Casale, would go far toward silencing his critics. With Louis XIII deeply engaged in the siege of La Rochelle, the chances of French intervention in Italy seemed remote, and the prospects for a brilliant Spanish success correspondingly good.

But although there seemed a logic to it, the intervention in Mantua proved to be a terrible miscalculation. The three-year struggle for the control of Mantua—which, in the end, was unsuccessful—placed an enormous additional strain on Spanish resources at a moment when the capture of the treasure fleet by the Dutch had brought temporary disaster to the royal finances, and when the Castilian economy, af-fected by poor harvests and rising food prices, was entering a new phase of recession. Worst of all, by dividing Spanish resources, it forced Spain to reduce its pressure on the Dutch, played havoc with Olivares's timetable for bringing peace to northern Europe, and set the monarchy on a collision course with the France of Richelieu.

Olivares, for all his bravado, was on the whole a cautious statesman on the world stage, and the Mantuan affair was out of keeping with his usual penchant for temporization. The gamble that he took on this occasion is not, I think, uncharacteristic of what can happen to the foreign policies of empires in decline. The tensions build up as they discover the limits to their power and fear the loss of face; internal difficulties begin to multiply; and then, in a moment of desperation, caution is thrown to the winds, and they seek to save themselves by the dramatic throw of the dice. If the Spanish army of Milan had succeeded in capturing Casale, the coup would have been hailed as a spectacular success, and the count-duke, with all the prestige of victo-ry behind him, might have turned with renewed assurance to his program for domestic reform. As it was, he made the worst of every world. The peace which he desperately needed was once again postponed, and with it his reforms. Until Mantua, he had attempted to run in double harness a policy directed toward reformation at home and reputation abroad, but after the Mantuan debacle, everything had to be subordinated to preparations for the anticipated war with

France. Already in 1632, three years before it broke out, the secretary to the Tuscan ambassador in Madrid reported that foreign affairs now took precedence over everything else.[20] Olivares would dispute the assertions of those who claimed that reform was impossible in wartime, and occasionally would attempt to breathe new life into his reform program, but in fact the only reforms that could be attempted in the 1630s were those that were directly related to the more effective conduct of war.

The Mantuan affair illustrates, I believe, the extreme difficulties of disengagement for an imperial power. The sheer extent of its commitments means that almost everything is perceived as affecting its vital interests. Hence the prevalence of the domino theory in the Madrid of the 1630s. But it is legitimate to ask whether anyone in the Spain of Olivares advocated an alternative foreign policy—one that would seek to reduce the area of its vital interests and, if necessary, sacrifice reputation to solvency. In other words, did anyone dare to think the unthinkable, the possibility of a staged retreat from empire? The nearest Olivares himself ever came to this was in 1633, when he at least played with the idea of a partial Spanish withdrawal from the Netherlands, knowing full well that simultaneous war against France and the Dutch could well bring the downfall of the Spanish monarchy; but in general he gives the impression of being unable to reconsider the assumptions that had traditionally governed the foreign policy of the Spanish Habsburgs.

Among Olivares's rivals and enemies I have found no trace of such an alternative foreign policy, although they were quick, of course, to denounce his intervention in Italy once it had visibly failed. It is also striking that, after his fall from power in 1643, there was no change of direction in a foreign policy to which the king himself remained firmly committed. But the degree of government censorship in seventeenth-century Spain makes it difficult to know what was really being thought, rather than expressed. There was certainly a continuing resentment against the sinking of Castile's precious resources, year in and year out, into the apparently bottomless pit of the Netherlands, but this resentment never found adequate expression either in the Cortes of Castile or in the ranks of the administration. The only document I have found for this period which suggests a radical rethinking of Spanish foreign policy was produced in 1635, three months before his death, by a friend and confidant of Olivares, the count of Humanes. In this paper he argued that Spain was in its present sorry condition because of its persistent determination to cling to Flanders and Milan. He therefore proposed that Spain should abandon the Netherlands and use the five or six million ducats a year that this would save to improve the defenses of the Iberian peninsula and the Indies. He also advocated that Milan should be ceded to the brother of Philip IV, the cardinal-infante, as an independent duchy. The monarchy, thus reduced to Spain, the Indies, Naples, and Sicily, would then be easily defensible, and the king would in reality be much more powerful than he found himself at present.[21]

Humanes's proposals, to my knowledge, found no echo among the councils in Madrid. Instead, the determination was to yield nothing, and fight it out to the end. But Olivares knew that, in doing this, he was engaged in a race against time. When war with France finally came in the spring of 1635, he was well aware that the only way to win was to win quickly. Spain was in no condition to withstand a war of attrition. It was for this reason that, within a few weeks of the outbreak of hostilities with France, he drew up a plan of campaign for a massive three-pronged assault on France from Flanders, the Austrian empire, and across the Pyrenees, in the hope of bringing the war to a speedy conclusion. Either, he wrote, everything would be lost, or else the ship would be saved. Either Castile would have to bow its head to the heretics, "which is what I consider the French to be," or else it would become "head of the world, as it is already head of Your Majesty's monarchy."[22]

With failure of the Corbie campaign of 1636, the conflict settled down into the kind of war of attrition which Olivares had been so anxious to avoid. The indecisive struggle of the later 1630s imposed growing strains, not only on the fabric of Castilian society, but also on the fabric of the Spanish monarchy itself, as Olivares sought to draw the various kingdoms and provinces into his Union of Arms. It was this growing pressure of state power on the peripheral provinces of the Iberian peninsula which provoked in 1640 those two great upheavals that were to have such profound consequences for Spain's international position: the revolts of Catalonia and Portugal.

The 1640 revolts suggested that the Spanish monarchy was on the verge of dissolution, and gave corresponding encouragement to Spain's enemies. "The greatness of this Monarchy is near to an end," as the British ambassador in Madrid observed.[23] In practice, reports of its demise proved premature. Olivares himself managed to hang on to power for a further two years, in spite of the calamities of 1640, and did something to contain, if not repair, the damage. He even succeeded by six weeks in outlasting Richelieu, whose death in December 1642—especially when followed by that of Louis XIII six months later—offered new hope to Madrid just when it seemed that all hope was lost. But the two revolts of 1640 inevitably had a major impact on Spain's military effectiveness in Europe, and forced the ministers in Madrid to reassess the international situation. Olivares himself, in what may have been his last state paper before his dismissal, seized on the news of Richelieu's death as providing the occasion for a new peace initiative. The news, he wrote, "compels His Majesty's ministers to consider the acute situation in which we find ourselves, so as to miss no opportunity afforded by this event to secure by any possible means a treaty of peace, which is all that can possibly restore our fortunes at this present juncture."[24]

But it is indicative, both of the tenacity with which Olivares's successors held to the guiding lines of his foreign policy, and also of the reserves of Spanish power and determination in the face of a succession of disasters, that the war with the Dutch would continue for a further five years after Olivares's fall, and the war with France for another sixteen. Even in retreat, reputation remained paramount. But, at the end, even reputation was largely lost. With the Peace of the Pyrenees in 1659, European primacy effectively passed from the Spain of Philip IV to the France of Louis XIV, and the Spain of Philip's successor, Carlos II, unreformed and languishing, became notoriously the sick man of Europe. Ironically, the Spanish monarchy that emerged from the peace settlement of Utrecht in 1713–1715, following the War of the Spanish Succession, was not unlike the one envisaged by the count of Humanes in his memorandum of 1635—a monarchy shorn of the Netherlands and its Italian inheritances, and, as a result, as Humanes had argued, a more viable monarchy with far better prospects for revival.

Conclusions

Looking back, then, over Olivares's twenty-two years in office, we see a signal failure to "manage" decline. Indeed, as I have suggested, Olivares did not even make the attempt. His intention was not to manage, but to halt, decline. In this he failed, and failed disastrously, as we, who know the end of the story, are well aware. But, at least until 1639–1640, it was by no means clear to contemporaries that things would turn out the way they did. They were more likely to be impressed by the extraordinary reserves of Spanish power, by the amazing capacity for recovery displayed by Spain in the wake of its reverses, and by the resourcefulness and energy with which the count-duke drove his flagging country forward. There were moments, like 1625 and 1634, when it looked as though Spain had the world at its feet; and, even after the outbreak of war with France in 1635, when Spain was fighting on two fronts simultaneously, it was by no means certain that—of the two great powers—Spain would be the first to collapse. If, for instance, the cardinal-infante had managed to reach Paris during the Corbie campaign of 1636, events might have taken a very different turn, and we might not be writing off the count-duke as one of history's great failures.

And yet, for all the count-duke's intelligence and resourcefulness, one is bound to ask whether, in the long run, the story *could* have ended very differently. In the late 1620s, when critics were commenting on Spain's mounting troubles and questioning Olivares's abilities, one of his colleagues remarked: "It is true that the end is approaching, but in other hands it would have come sooner."[25] There was a perceptible mood of defeatism in the Madrid of the 1620s and 1630s, a mood against which Olivares struggled but which, in his heart of hearts, I believe he shared. I suspect that the very concept of decline, with its sense of inevitability, contributed subtly to the creation of this mood, as also did the sense that a direct relationship existed between national and personal morality, and national success. As failures multiplied, Philip IV and Olivares attributed them to the sins of the nation, and above all their own. Like some latter-day Jonah, the count-duke in the wake of some defeat would say that the storm would not be stilled until they threw him overboard.[26] This is hardly the mentality that commands success.

But the pessimism also derived from what were in fact the very real, and growing, difficulties created by the attempt to hold fast to the traditional aspirations and ideals of the Spanish monarchy at a time when resources were shrinking, and the problems of imperial defense multiplying as a result of changing power relationships. As contemporaries like Botero recognized, a dispersed and global empire, made up of a wide variety of territories, was bound to be at a disadvantage in a conflict with more geographically compact states, whose resources were more easily mobilized and where power was concentrated.[27] Olivares was well aware of this, and took steps to meet the challenge by devising his Union of Arms. But a scheme like the Union of Arms was probably beyond the political and logistical possibilities of seventeenth-century societies, and Olivares found himself struggling to reform structures ossified by time and resistant to change.

The only remaining alternative, as I have suggested, was to reduce the level of aspiration and cut the commitments. But this was to turn one's back on a glorious past, and ultimately to deny everything for which the monarchy had stood. Spain was a society forged, and made great, by war, and its criteria for judging success or failure were military criteria. Decline was therefore measured above all by the loss of international standing and military power. For Olivares, as for most of his colleagues, the burden of the past was too heavy, and the price of voluntary retreat too high a price to pay. If it proved impossible, as seemed likely, to reach harbor safely, then, rather than surrender, it would be better to go down with the ship. It was necessary, as the count-duke said on one occasion, to "die doing something."[28] This was his ultimate grand strategy, and he proved as good as his word.

7

Total War for Limited Objectives: An Interpretation of German Grand Strategy

Dennis E. Showalter

The military heritage of modern Germany is a paradox. On one hand, the armed forces are presented as committed to preserving traditional sociopolitical positions in a modernizing society, increasingly concentrating on privilege at the expense of efficiency, ultimately succumbing body and soul to national socialism in a desperate effort to sustain an inflated self-image as the ultimate bearer of national identity.[1] On the other hand, the German military's fighting capacities are consistently praised—most of all by Germany's enemies. Evaluations of the army's operational performance in World War I are generally high. Words like "remarkable" dot the pages of professional and academic accounts, even those otherwise critical of the Second Reich's military effectiveness.[2] Moving to a more recent era, scholars like Martin van Creveld, journalists like Max Hastings, and statisticians like Trevor Dupuy are at significant pains to demonstrate not merely the Wehrmacht's skill at arms, but the superiority of its fighting power to its U.S., British, and Soviet enemies.[3]

Attempts to resolve these dichotomies all too frequently bog down in mutual accusations of tunnel vision, careless scholarship, and bad faith. Yet there is one significant point on which parties to the debate are generally able to agree. Whatever "genius for war" may be incorporated in the German military tradition does not extend to the realm of grand strategy. From the turn-of-the century days of Count Alfred von Schlieffen, and arguably before, German strategic thought has allegedly devolved downward, toward the tactical and operational levels, rather than upward. The German military establishment developed plans to win campaigns, not wars. It ignored or misread the signposts indicating an era of total mobilization, of comprehensive national commitment to a war effort. Germany's soldiers substituted cleverness and willpower for the objective, material factors that became the ultimate determinants of victory, at least in a prenuclear environment.[4]

To write of "soldiers" is a conscious choice. German grand strategy was ultimately a military strategy. The navy may have developed its own concepts, but like the fleets that implemented them, they are best described as luxuries. As for the air force, what Germans call *das Gesetz des Handelns* impelled it almost from its beginnings away from any pretensions to an independent role. Germany's location made her ground forces a sine qua non of her existence and her status. It is there that analysis promises the greatest profit.

This essay begins with an assumption, no less important for being banal. It accepts as given that German soldiers did not prepare for wars they expected to lose. It also hypothesizes that the average General Staff officer was no less perceptive than the average defense analyst or historian. In other words, German military planning incorporated reasoned judgments of German military requirements and prospects. It might stress the balance between means and ends to the limit, but never systematically exceeded them as a first choice.

To understand the nature of German strategic concepts, it is also necessary to take a broader historical approach than current concentration on the Second Reich and the Wehrmacht makes fashionable. A crucial key to the German way of war is a Prussian tradition significantly concerned not with unchaining Bellona, but with taming and retaming that fickle goddess. Prussia's eighteenth-century status as an intermediate power in an environment which did not wish her well fostered a pattern of waging general wars for limited objectives. The reign of Frederick the Great in particular had been characterized by the escalation of policy conflicts into struggles for the state's existence, and by the corresponding evolution of an approach to strategy that sought to avert that contingency by emphasizing prompt, decisive defeat of an enemy as a prelude to a negotiated peace. Apart from the long-term risks of domestic disintegration through protracted conflict, indicated in the later stages of the Seven Years' War, logistic factors precluded concentrating large forces on Prussia's frontiers for any length of time.[5]

The experience of the revolutionary/Napoleonic era further enhanced Prussia's concern with limitation. French systems and attitudes, based as they were on constant change and improvisation,
required institutionalizing if they were to be adapted to the needs of a
state that proposed to form part of a stable European system. On the
operational level, battles became increasingly indecisive as armies
grew too large and too unskilled to do much more than grapple in-
coherently with each other. The resulting extension of conflict was
perceived as destabilizing existing social and political institutions
sufficiently to encourage conservative Prussian military theorists to
focus during the first half of the nineteenth century on the problem of
ending any future wars as quickly as possible.[6] Even Clausewitz, in
arguing for the importance of finding an enemy's "center of gravity,"
concentrated on armed forces, paying at best parenthetical attention
to such general factors as alliances, public opinion, and domestic
stability.[7] Economic considerations played no significant role in his
writings. And when the Prussian army did begin paying attention to
the Industrial Revolution, it was within the framework of operational
considerations.[8]

The success of this approach seemed plainly manifested in the Wars
of Unification. However revolutionary the results of these conflicts
appear to observers and historians alike, it is necessary to remember
that Prussia's grand-strategic goals were ultimately negative and ultimately limited. Like the American Revolution, the German Revolution of 1866–1871 was more turnout than turnover, involving Austria's expulsion as much as Prussia's expansion.[9] The organization of
the North German Confederation and the Second Reich with significant federal dimensions further eased the transition for many of the
small and middle-sized states, whose inhabitants and officials alike
would have resisted far more strongly their simple absorption in a
Prussia written large.[10]

Prussia's experience both reinforced and reflected Chief of Staff
Helmuth von Moltke's evolving notion of a logical and necessary
demarcation between the spheres of war and diplomacy. Moltke's
insistence that war once begun must be left to the generals was accompanied—at least prior to 1870–1871—by a conviction that in practice, an enemy decisively defeated would be amenable to negotiations. Then, by Moltke's own logic, the statesman must take over. It
was the army's task to bring about the conditions for peace, as opposed to destroying directly a foe's entire capacity for resistance.
More than many of his counterparts and successors, Moltke recognized the latter task essentially impossible. Even under the condi-

tions of total war in the Napoleonic pattern, states and societies retained even in defeat a significant ability to fight back. Spain of the *guerrilla,* Prussia of the *Befreiungskrieg,* were only the most obvious examples, to say nothing of the Confederate States of America. Responsible use of armed forces correspondingly involved convincing an adversary not to use that capacity further.

This requirement was further strengthened by Moltke's acceptance of the dictum that war is the province of confusion, with no plan surviving initial contact with the enemy. That contact, therefore, must be made to count. Prewar preparation must minimize or eliminate wartime confusion. In this context the American Civil War and, to a lesser extent, the people's war of the French Third Republic represented departures from a norm. Moltke, far from dismissing the American experience, interpreted it in terms made familiar by Emory Upton. A national government lacking the force to suppress rebellion at its outset had been forced instead to improvise. The resulting war of attrition was predictable, but hardly desirable. The Franco-Prussian War manifested a similar pattern. The German armies, for all their initial successes, were neither strong enough nor smart enough to crush the improvised forces of the Republic before their existence prolonged the conflict to a point seriously threatening Germany's European position.[11]

What conclusions might be drawn from such data? Strategy is best understood as the calculation of relationships among means, ends, and will. Let this process break down in any area and the result is not bad strategy, but no strategy. Germany's strategic doctrines in the last quarter of the nineteenth century developed in the context of an emerging *Weltpolitik.* Critics of the Second Reich consistently point to the increasing contradictions of its foreign policy even under Bismarck. Germany's vital interests on the European continent demanded stability; Weltpolitik was in its essence a policy of confrontation. The dichotomy forced Germany into a galvanic stasis—any movement in one direction was checked by a shock from another. Initiatives in the Far East or the Ottoman Empire were inextricably interlocked with events on the Rhine and the Vistula.[12]

This contradiction is obvious enough to invite interpreting the Second Reich's foreign relations as manifesting the tensions of a domestic structure whose malaise encouraged, not to say demanded, the outward projection of national power as a guarantor of internal stability.[13] This by now familiar concept of Germany as a semimodern state whose bumptiousness imperfectly masked a sense of weakness overlooks many things that were right with the Reich, and many others that were perceived as right. With its central strategic location and burgeoning economy, with a strong monarchical system and a vocal, influential parliament, Germany in the early twentieth century saw itself as cut off from none of its neighbors, able to associate with any of them.[14] This generated a corresponding desire, whether in the ambitious versions of the period of Bernhard von Bülow's chancellorship (1900–1909) or the modified visions of Theodor von Bethmann Hollweg, to establish Germany as the center of Europe's diplomatic network. The Second Empire would simultaneously contain Russia's aspirations and integrate Britain into the Continental balance of power, all the time herself remaining firmly bound to nothing except her own choices.[15]

The risks of such a policy, the need to back it with superior, if not overwhelming, military force, appear obvious at this distance. Even to contemporaries German behavior generated enough anxiety to foster a series of negative combinations against a Teutonic threat exacerbated by Kaiser Wilhelm's erratically bellicose rhetoric. By 1908 at the latest, encirclement, *Einkreisung*, had evolved from worst-contingency nightmare into everyday reality.[16] Yet Germany, like Europe's other great powers, never became consequently belligerent. The significant constant in Wilhelmine foreign policy is its reluctance to translate the rhetoric of violence into action. The Reich's repeated crises manifest a similar pattern, with newspaper headlines, parliamentary speeches, and foreign office memos resembling a fever chart. Initial stirrings zigzag to a peak of martial enthusiasm, followed by a sudden, sharp drop in belligerence just as war hysteria seems on the verge of explosion.[17]

Gunpowder was, in short, a potent antidote to chauvinism—so potent that a perceptive observer like Kurt Riezler incorporated it into his advocacy of brinkmanship as a tool of foreign policy. If Germany held on long enough in a crisis, he averred, her adversaries would blink first.[18] It was the kind of reasoning that carried more force in the abstract than in actual situations. Certainly Germany's decision makers, military and civilian, were sufficiently aware of the lability of their domestic support to be reluctant to accept head-to-head confrontations as a diplomatic norm. The popular enthusiasm that sustained World War I was in fact a post facto phenomenon, as surprising as it was gratifying to a power structure which expected to face significant internal opposition to war waged for any reason.[19]

The contradictions of German foreign policy combined with the instability of German domestic politics to put massive demands on the German military establishment. Its tradition, its structure, and its

matrix emphasized quick, decisive victories as a prelude to negotiations. But such victories in the context of the Reich's increasingly
ambitious diplomatic vision were perceived as difficult to achieve.
Even before the emergence of Alfred von Tirpitz, naval planning
against any likely combination of enemies had an Alice in Won-
derland quality, a casual acceptance of contradictions and if-then
contingencies foreign to the Prussian tradition. An Admiralty Staff
that denied any chance of war with a single great power found itself
simultaneously warning the kaiser that eagerness for battle did not
mean risking suicide against superior forces.[20] Whether the fleet
should be used at the start of a war, or used at all, was a complex
question depending on specific circumstances. Exactly how those
circumstances could be adjusted to Germany's advantage remained
the stuff of dreams and speculations. A strong case can be made that
the fleet's very existence nurtured a climate of unreality that boded no
good for German grand strategy in general.[21]

The army faced similar problems. In the aftermath of the Franco-
Prussian War, Moltke's worst-case contingency, a war on two fronts
against France and Russia, involved ripostes in both theaters with the
aim of forcing at least Russia into negotiations. By the 1880s, the
growing spectrum and depth of Russia's germanophobia combined
with the military and economic recovery of France to make that contingency at best a high-risk gamble. At the end of his career Moltke
had moved beyond considering the option of preventive war and was
openly questioning the prospects of using armed force to implement
state policy under any conditions.[22]

Like most of his counterparts and successors, Moltke feared the
implications of modern war. Nationalism had intensified its emotions; industrialization had exacerbated its destructiveness. Any system subject to its rigors for any length of time was likely to collapse
from physical and psychic stress, but Germany's fate in this context
seemed virtually certain to her military leaders. The Second Reich
was at once too divided against itself and too finely tuned to stand
much external pressure. Whatever strides Germany might be making
toward parliamentarization, *Sammlungspolitik** and grand coalitions alike had failed to produce a viable political synthesis.[23] Social
democracy remained imprisoned behind the "35 per cent barrier," but
within that limitation was consolidating its voters under the banners
of an ideology whose principles and rhetoric continued to stress revo-

*Literally, "grouping policy," or that of "rallying" progovernment forces in sup-
port of the Reich's policies.

lutionary militance.[24] Talk from the right of coups d' état and whiffs of grapeshot foundered on the social and political realities of a complex industrial society. Neither the War Ministry nor the General Staff regarded this as fertile ground for a deliberate policy of *Flucht nach Vorne*, of balancing domestic books by foreign war. For any doubters the Russian experience in 1904–1905 offered a plain object lesson of the risks such an approach entailed.[25]

In addition to these internal factors, the strong points of Germany's major likely enemies also suggested the impossibility of successfully waging sustained war. Great Britain's empire and, increasingly, Russia's economic potential were cards best played in a long game. Russia, moreover, was able to project her power directly at Germany's most vulnerable point: her eastern border. Even minor improvements in the quality and viability of her army meant an exponential increase in the threat not only to Germany itself, but to the pattern of limited wars that had created and sustained Prussia/Germany as a great power.[26]

The German military thus defined its task in the context of the destructiveness of modern war and the long-term strengths of Germany's enemies. These factors made it correspondingly necessary for the Second Reich to maintain the strategic initiative, to lead from its particular strengths in order to fight the kind of war Germany could win—a war kept within temporal and material limits.

This requirement generated two basic decisions. The first involved developing the German army as a rapier, not a bludgeon. If the empire conscripted fewer and fewer of its eligible sons between 1871 and 1913, this reflected more than a fear of corrupting the rank and file with Marxists and diluting the officer corps with social undesirables.[27] Schlieffen in particular recognized the ineradicable imbalance in manpower between Germany on one hand, France and Russia on the other. Even by training every fit man, Germany could not hope to match her enemies numerically. Individually, German soldiers might be the best in Europe. But German observers did not accept uncritically the biologically based nationalist criteria so influential in fin-de-siècle public opinion. The German soldier, they argued, was difficult to drive like a Russian, and even more difficult to exalt like a Frenchman. When brought to emotional heights, his response was likely to be a berserker's fury, at best marginally useful on the modern battlefield. Training, discipline, motivation—these things were helpful. But in an age when all armies were trained, armed, and equipped essentially alike, the prospects for securing more than a marginal advantage in quality seemed severely limited.[28]

Political concerns, domestic and international, also militated against drastic increases in the army's size. Germany's victories in 1866 and 1870–1871 had established her beyond any question as the Continent's premier land power. Anything more than incremental increases in her military potential were likely to be perceived by her neighbors—including Austria-Hungary—as a direct threat calling for an immediate response. This kind of arms race was the last thing Germany needed or wanted. Not only would it enhance the strain on the foreign office, it would put pressure on a military budget whose approval by the Reichstag was by no means an automatic process. Especially after Bismarck's fall, an increasingly bureaucratized chancery was reluctant to spend any more of its political capital than absolutely necessary in debating parliamentarians with free access to the media.[29]

The focal point of these attitudes within the army was the War Ministry, whose own increasing politicization gave it broader vision than its great institutional rivals, the Military Cabinet and the General Staff.[30] Karl von Einem, who held the post of war minister from 1903 to 1909, waxed scathing about a General Staff that was in the comfortable position of making proposals in the abstract, while his office had the responsibility of translating theory into practice.[31] Even General Erich Ludendorff was bearded to his face by a colleague in the War Ministry who bluntly informed him that his seemingly limitless demands for more would drive the German people to rebellion.[32]

These and similar observations were not lost on a General Staff whose increasing specialization encouraged respect, however grudging, for other agencies' specialized competence. Wilhelm Groener, one of the prototypes of the up-to-date General Staff officer, was to comment ruefully that national economics was not part of the curriculum in officer training.[33] The War Ministry correspondingly tended to win a lion's share of the arguments over force structures. Given the resulting limitations on numbers, the General Staff coped with Germany's strategic liabilities by emphasizing its own perceived best quality: professional skill at higher levels of planning and command. Instead of playing to its adversaries' strengths by a series of head-to-head encounter battles, the German army must seek to change the rules, to impose a plan so comprehensive, so cohesive, that the enemy would be able to do nothing except react. Far from ignoring or denigrating the power of modern weapons, Schlieffen proposed to take advantage of them as force multipliers, reducing the strengths of covering and screening forces to what seemed an unacceptable minimum to more conservative colleagues. But the General Staff exercises of the

1890s in both east and west indicated the possibilities even under modern conditions of a small force overcoming a larger one by concentrating against the enemy's flank, then striking against his lines of retreat.[34]

This principle was best applied against the one adversary whose defeat no German soldier doubted: France. This attitude did not mean the Germans despised the French as an enemy. But since 1870 the French army had essentially formed itself according to patterns set across the Rhine. Despite specific advantages in some areas, it continued to be viewed as a blurred copy of its original.[35] Aside from the advantage conferred by Germany's larger population, the Second Reich's military planners were convinced that France could be beaten man for man and corps for corps. The growing faith among Europe's military planners in the tactical and operational superiority of the offensive only strengthened the conviction that an all-out attack on France meant removing not only an immediately dangerous enemy, but the one most vulnerable to a Germany herself in no position to sustain a long-drawn-out war.[36] The Schlieffen plan, far from being an infallible recipe for victory, was a least-worst solution, proof of the homely axiom that when one is handed lemons by life, the best response is an attempt to make lemonade. In psychological terms, the plan offered hope through diligence. If everyone did his part superbly, if every element of the machine performed according to expectations, Germany's window of vulnerability might become a door of opportunity.

Schlieffen's approach made operational planning the key to grand strategy. In January 1914 a conference, including representatives from the War Ministry and the General Staff, and the chiefs of staff of all twenty-five army corps, was held in Frankfurt am Main. Its theme was mobilization—how to make the process of shifting from peace to war footing faster and more efficient. There was no talk of harnessing the nation's entire resources for a life-and-death struggle. Instead the War Ministry counseled against suspending the publication of Social Democratic newspapers until they behaved "in a way hostile to the fatherland." There was no need to drive them into the enemy camp. The conference criticized "on social grounds" the navy's practice of conscripting more men than it required, then discharging the surplus. Victims of this practice usually found their jobs filled when they returned home, with corresponding negative effects on their enthusiasm for military service.[37]

Anything further from detailed planning for total war can scarcely be imagined. But the nature of the meeting reflected the fact that for

German military planners, victory meant less the destruction of the Entente than its neutralization. It meant breaking not the capacity to resist, but the will to use that capacity—or what remained of it after the "battles of annihilation" the German army expected to force in the war's opening weeks. Germany prior to World War I did not see herself bestriding the ruins of former enemies. She sought preeminence in Europe, not hegemony over it. The difference is subtle, but not artificial. And its ultimate manifestation was the continuing acceptance by the military of the principle of negotiated peace as the logical consequence of victory on the battlefield.[38]

This also meant accepting the ultimate supremacy of Germany's political authorities. The dichotomy established between Moltke and Bismarck in the 1860s survived another half-century. By 1914 it had been intensified on one hand by the perceived necessity for the military to concentrate on an ever-narrowing field of professional, technical responsibilities; and on the other by a growing indifference to the details of warmaking in the Chancery, the foreign office, and the Reichstag alike. The army's increasingly jealous guarding of its turf was underwritten by officials and politicians all too willing to assume the soldiers knew their own business, and to accept the Reich's armed forces as competent by definition. The generals for their part thought no further than delivering the next war's enemies on a silver platter, doing their half of the job as thoroughly as possible in order to give the men in morning coats as little as possible to spoil. Throughout World War I their perceptions would feature a significant dichotomy between the relatively sober evaluation of operational possibilities and the wish-dreams of their political and economic visions.[39]

This fragile interlocking structure of specializations collapsed as the guns of August 1914 switched from preparing German attacks to bombarding Allied trenches. The ultimate problem of World War I was neither strategic nor operational, but tactical: covering the proverbial last three hundred yards to an objective, with enough resources remaining to pursue immediate victories.[40] Not until 1918 would any of the combatants successfully solve the overlapping challenges posed by machine guns, barbed wire, and internal combustion engines still in the beginning stages of development. In the meantime Germany found herself trapped in exactly the worst-case contingency her military planners had sought to avoid: a two-front war of attrition. No combination of force and diplomacy proved able to break the deadlock.

Deadlock threatened to become gridlock after 1916. The drastic enlargement of military authority, making Ludendorff and Hindenburg the virtual dictators of Germany, was not accompanied by a

corresponding increase in military power. If there was little direct
opposition on the home front to the army steamroller, the soldiers
nevertheless found themselves responsible for dealing with many of
the same problems that had proved intractable to civil authorities for
months and decades. They made a royal hash of their opportunity.
Military censors found themselves outwitted on a daily basis by
newspaper editors tempered under the Reich's anti-Socialist legisla-
tion. Military authorities bore public responsibility for the inconsis-
tencies of a rationing system that allowed those with money or connections to live much as they pleased. Instead of giving orders,
Ludendorff and his associates found themselves negotiating with
unions and businessmen alike to keep Germany's overstrained industries from self-destructing. And if the military dictatorship was more
or less successful in breaking strikes, it proved completely incapable
of checking rampant profiteering. In the biting words of Matthias
Erzberger, after three years of struggle the War Ministry had shown
itself powerless against the Daimler Company.[41]

Much of the army's prestige in the Second Reich had depended on
the limited demands it placed on society at large. In training camps
and replacement depots, a new generation of conscripts was increasingly reluctant to turn away from the clandestine propaganda
describing them as cannon fodder for the big shots. Among civilians it
did the military's image no good to be perceived as the enforcer of
petty regulations, confiscator of the dozen potatoes and the jar of fat
that were the painful fruits of a day's illegal scrounging in the
countryside.[42]

The soldiers responded on two levels. Relatively junior officers like
Colonel Max Bauer attacked democracy, feminism, constitutionalism,
socialism, and Jews in a bewildering farrago of ideas that has been
described both as proto-fascism and as an extension of views already
common in military circles. It remains questionable, however, which
was the chicken and which the egg—whether the war was ultimately
seen as a means of suppressing domestic reform efforts, or whether
that position gained respectability and credibility as traditional con-
servatism lost ground in the eighteen months before the empire collapsed.[43] For a long time in 1917 members of the active armed forces,
legally forbidden to belong to any political party, were allowed, if not
exactly encouraged, to join the right-wing Vaterlandspartei on the
grounds that it was a patriotic movement rather than an organized
party. This rapidly proved too tame. By November 1918, some extrem-
ists were discussing the best ways of employing tanks and aircraft
against "Bolsheviks" on the home front.[44]

Such posturing was only one side of the story. From the War Ministry downward warnings were raised of the risks of politicizing the army from any direction.[45] Ludendorff might represent the wave of the future to an emerging radical right. To many of his colleagues he was a Pied Piper—the same man deemed dangerous enough in 1913 to be prematurely transferred from the General Staff to a line regiment.[46] And those who favored, or came to favor, total war as a means of reconstructing Germany dared not overlook the sine qua non: victory in the field.

This knowledge was more widespread the nearer one got to the front. In terms of operations, Germany's military leaders remained convinced that a bit more of the same would bring ultimate victory— or at least a favorable decision. Their attitude was more than the vitalism described by C. S. Forester as the military equivalent of trying to pull a screw from a block of wood by direct force. Experience on the eastern front exercised a significant influence. Since Tannenberg, Ludendorff had pursued in Russia the fata morgana of a victory that always seemed just within his grasp—if only he had a few more divisions, a bit more ammunition, a slightly stronger ally. The quartermastergeneral brought this mindset to Berlin. It was strengthened by the significant improvements in tactical doctrine and infantry-artillery cooperation during 1917, and by exposure to the navy's stubbornly tactical orientation in pursuing the submarine campaign. By the time of the great western offensives in March 1918, Ludendorff had reached a point where his thinking was not even operational in conception. Let us punch a hole, he argued. Let us first defeat the British and the French. All else will follow from that.[47]

The approach was not intrinsically feckless. It incorporated rather the wisdom of the recipe for rabbit stew that begins: "First, catch one large, plump rabbit." Ludendorff recognized, more clearly than did many of his future critics, the essentially tactical nature of World War I. He also recognized the German army's weaknesses as an instrument of exploitation. By 1918 most of the cavalry had been dismounted. Armored cars were an expensive luxury under conditions prevailing in the west. In 1914 Germany's Jäger battalions made good use of their organic truck columns, while cyclist and motorized units were successfully improvised in the eastern theater. Two years later an improvised battle group based on a battalion of truck-mounted infantry had played a key role in the Rumanian campaign. But the prospects for operational deep penetration against a determined enemy under the conditions of the western front were too remote to merit consideration.[48] What Ludendorff proposed instead was to break his enemy's will by repeated tactical hammer blows. A century and a half of conditioning led him to believe the inevitable next step would be a negotiated peace—negotiated on German terms and at the sword's point, but negotiated nevertheless. Germany's resources, however, were no longer equal to the demands of a sustained offensive, no matter what its ultimate objectives, or lack of them. Ludendorff's poorly used tool finally broke in his hands.[49]

The collapse of the Second Empire encouraged reappraisal of Germany's military heritage. Defeat in the world war was almost universally interpreted among the military as a result of half measures. Imperial Germany was criticized for failing to integrate the moral resources of its society, the economic potential of its agriculture and industry, and the material resources of its allied and occupied territory into the war effort. In the immediate postwar years elements of the military establishment developed a significant internal dynamic, encouraged by an emerging civilian radical right, supporting the continuance of the Hindenburg/Ludendorff dictatorship—if necessary under the auspices of a Social Democratic civilian, War Minister Gustav Noske.[50]

The limited appeal of this policy inside and outside the army did not mean that it disappeared after 1923. The Weimar Republic was characterized by systematic extension of state intervention and government control in the name of modernization. The Reichswehr was hardly exceptional as it sought to expand its influence with an armaments industry able, in the context of Versailles, to do little but dream of future orders. It also extended and deepened its contacts with the foreign office. Pan-German visions of an empire in the East were overlaid by consideration of Russia's role in any future German war economy, whether as a supplier or a satrapy.[51]

The tendency to make both foreign and domestic policy into the conduct of war by other means was fostered, not to say required, by a peace treaty that had left Germany defenseless in any conventional sense. Whatever the Reichswehr's specific strengths and weaknesses vis-à-vis specific potential enemies, in practical terms Germany could not count on a military solution to her security problem.[52] This in turn encouraged her to move outside traditional perspectives. Successive Weimar cabinets depended increasingly on economics as the best means of securing that revision of Versailles called for in party programs far into the left of the spectrum.[53] A related approach to national security, particularly congenial to soldiers deeply involved in politics like Wilhelm Groener or Kurt von Schleicher, argued for blending diplomacy and military planning to develop the forerunner of a European deterrence system based on a general distaste for war.

German rearmament, these men suggested, could not take place in a vacuum because of Germany's current disarmed status. She was so far behind in the race that her potential rivals could stay ahead of her with only limited efforts. What was needed instead was a policy encouraging Germany's neighbors to accept rearmament as, ultimately, a stabilizing force and a first step to revising Versailles by consent.

An alternate concept proposed the autonomous development of German security policies. Its advocates thought in defensive terms: unleashing a people's war against any invasion. Their line of argument, however, had significant limitations. Apart from the obvious difficulties of mobilizing the fragmented society of Weimar Germany into a functioning guerrilla force, the remedy seemed worse than the disease. Insurgency meant destruction and disruption. In short, both negotiations and la guerrilla seemed unacceptably high-risk options in the context of a European system that continued to insist on the Versailles Treaty's disarmament clauses as the rest of the structure crumbled and collapsed.[54]

One result of this perceived dead end was that the military's initial contempt for and mistrust of national socialism increasingly gave way between 1930 and 1933 to an image of Hitler as a deus ex machina. The Nazi movement appeared to offer the mass base necessary both to facilitate national mobilization and to legitimate the armed forces' privileged status. Doubts about Hitler's willingness to settle for half a loaf diminished when Reichspräsident Paul von Hindenburg unilaterally appointed Werner von Blomberg as minister of defense. They faded into insignificance when Hitler swore, as a condition of his appointment as chancellor, to refrain from interfering in army matters. Particularly after the Blood Purge of 1934, when the army's benevolent neutrality facilitated Hitler's massacre of the SA's leadership, the soldiers appeared to have not merely sustained, but enhanced the dualism of military and political forces which had prevailed under the last years of the Prussian monarchy and the first years of the German empire. Even Nazi propaganda increasingly trumpeted the "two pillars theory," of a Reich built on the mutual relationship of party and army.[55]

Ultimately the army's relationship to national socialism reflected the limitations of the soldiers' political commitment. Their domestic position since the days of Frederick the Great had rested heavily and increasingly on their ability to win wars quickly and at limited cost. Their status in the Second Empire reflected the consent of opposing forces at least as much as any specific manipulative or coercive skills.[56] After the debacle of 1914–1918, Germany's military elites did not feel constrained to choose between their professionalism and their sociopolitical pretensions. Instead the latter was increasingly perceived as depending on the former. What the army *was* depended on what the army *did*. Its special role in German life was a function of its ability to aggrandize the German state and defend the German people.

This attitude must not be uncritically accepted as a comprehensive explanation of the military's behavior. Nevertheless, it is equally inappropriate to reject the soldiers' overt justifications as manifesting nothing more than mendacity or self-deception. The army's mindset, moreover, was hardly original, or unique to Germany. What did distinguish German military thought in the Weimar era was the fact that it developed in a vacuum. From Hans von Seeckt downward, planners dealt in terms of armed forces that did not exist. Balanced thinking was at a corresponding discount. Historically, a major strength of the Prussian/German approach to war had been its rootedness in the concrete. Force structures and strategies had been developed in response to specific circumstances of state and society. Now the restraints of Versailles lured German soldiers into the empire of dreams.[57]

A similar situation existed across the Atlantic Ocean, where a United States army barely adequate to perform a rapidly declining constabulary mission developed elaborate theoretical plans for national mobilization of human and material resources under military auspices. Here too the blatant discrepancy between existing means and probable missions encouraged concentration on total war to the exclusion of any alternatives.[58] Far more than in the U.S., however, in Germany the development of operational strategy became an end in itself, a process that discouraged serious thought about what happened next.[59] A major consequence of the Versailles Treaty was its contribution to encouraging in the German military an abstract approach to future war that increasingly made rearmament its own justification. Neither Reichswehr nor Wehrmacht had a clear concept of purposes or policies. Security was seen not in the relative terms of the eighteenth and nineteenth centuries, but in absolutes. In November 1938, Ludwig Beck soundly declared that on rereading Ludendorff's *Total War* he found himself not only agreeing with every word, but going further than the author.[60] The soldiers, like Faust, seemed to believe themselves damned only when they accepted limitations.

This flight into military metaphysics had significant consequences in the Third Reich. By 1938 the Wehrmacht's ill-conceived, poorly

coordinated arms procurement policies had overheated the economy
to the point of self-destruction—or the point where the direct absorption of additional resources was arguably necessary to avert an explosion.[61] The prospect was hardly daunting to an emerging generation
of technocrats. If Moltke and Schlieffen believed in the power of
planning to change circumstances, their successors placed even more
faith in weapons and methods. Germany's soldiers saw World War I as
having been shaped, more than most wars, certainly more than any
war in the Prussian/German experience, not by strategy, not by logistics, not even by operations, but by tactics: the difficulty of achieving
battlefield victory except at ruinous cost, and the impossibility of
converting that victory into a decisive breakthrough.[62] By the
mid-1930s things had changed. The concept of blitzkrieg was far less
developed than postwar popular mythologies concede. Nevertheless,
to the Becks, the Guderians, and the Mansteins, to the junior officers
who would rise to lead corps and divisions, the internal combustion
engine was not altering the parameters of strategy but restoring them.
Prospects of decisive victory, as opposed to bean-counting attrition,
seemed to be reemerging.[63]

But reemerging to what purposes? For what ends were the massive
armament programs, the new techniques and technologies, to be
used? Germany's material limits imposed a correspondingly strong
moral requirement to maintain that rational balance among means,
ends, and will described earlier as the essence of strategy. But this in
turn demanded a degree of detachment foreign to the modern state,
democratic or totalitarian, which fosters by its nature the integration
of every aspect of life into the political community. The detachment
that nurtures balance was particularly difficult to sustain in a German
military establishment that even before the Second Empire had begun
the process of identifying itself with national values and citizen-
formation.[64] The concepts of total war considered before and during
World War I and developed under the Weimar Republic were increasingly seen as requiring a synthesis of the armed forces with the
national community. Hitler's accession to power could correspondingly be presented as a matter of tactics rather than principle, an
acceptable means to a greater end.[65]

The Führer had another function as well. Since the days of Frederick the Great, the Prussian/German army had fought total wars for
limited aims, applying the maximum force possible within a framework established and controlled by the statesmen. In the context of
strategic policy-making, this was the role the army assigned to Hitler.
It was not entirely the product of self-deception. Hitler had some

sense of Germany's inability to sustain a long war, and more sense of the risks involved in waging war against a coalition. Between 1935 and 1938 he demonstrated an uncanny ability to utilize both the shadow and the substance of military force to achieve disproportionate diplomatic triumphs. After Munich, even critics of the Nazi regime wondered if the great days of Bismarck and Moltke had not come again. However Hitler's diplomatic high-wire acts might frighten his generals, he seemed to have a sixth sense for how much the traffic would bear in any given case. The army's cooperation was further justified by its own emphasis on the "two pillars theory" referred to earlier, which suggested that as long as the second "pillar" established the matrix of military success, the soldiers were best occupied in their own sphere.[66]

The steadily increasing erosion of the presumed symbiosis between soldiers and statesmen after 1938 went, if not exactly unremarked, certainly unacted upon.[67] This passivity was in good part a product of calculation. Between the Munich conference and the launching of Operation Barbarossa, most of the senior officers were able to justify themselves as doing things they and their predecessors had always done. They were preparing to fight and win wars quickly, with the nation's vital energies focused through the military system and under its control. Private armies like Goering's Luftwaffe ground troops, even the embryonic Waffen-SS, could be tolerated, even welcomed, in this context as gestures of goodwill toward a regime that, at least before 1941, ultimately seemed willing to trust the everyday conduct of its war to the professionals.[68] And all the army had ever felt it needed, operationally, was ultimate control of the last three hundred yards of the battlefield.

But Hitler refused to adapt to the army's matrix. For the first time in their history Germany's armed forces performed in the context of a system deliberately unlimited in its seeking of enemies. They served a regime whose ideologically based domestic and foreign policies denied the dualism of the Prussian/German army's historic approach to warfare.

German soldiers went to war in 1939 with little sense of a relationship between means whose limits they perceived all too clearly and ends which seemed to exist only in the mind of the Führer.[69] For twenty months, however, they functioned in a setting that seemed familiar—of negotiated peaces with defeated enemies. Franz Halder, Beck's successor as chief of staff, was able to record without comment even Hitler's repeated declarations of his intention to replace the Soviet Union with a network of German client states that would be at least nominally independent. It was a comforting fig leaf.[70] And when it no longer sufficed, there remained an ever-deeper withdrawal into a self-proclaimed "state of being only soldiers" (nur-Soldatentum), an ever more punctilious emphasis on the shadows of honor as its substance eroded. The military leaders took their marshals' batons, their grants of cash and land. They saw their colleagues villified, dismissed, ultimately hanged from meathooks, and all without significant protest.[71] But silence was only part of their integration. The relationship between the Reich and its soldiers featured positive cooperation as well, especially in the east, where the Wehrmacht's overt compliancy in Nazi crimes reflected an internalization of Nazi ideals that was all the greater for being unacknowledged and unacknowledgeable.[72]

It all availed nothing. Neither blindness of eye nor suppleness of spine kept the generals from losing control of even the army's most jealously guarded preserve. By war's end the last three hundred yards were so firmly under Hitler's control that his Commander in Chief West said that the only troops he was authorized to move were his headquarters sentries!

This discussion of German grand strategy ends as it began: with a paradox. But it is a different kind of paradox. The military's vitalist approach to national strategy was more than a foredoomed effort to square the circle, to cope with an inherently unstable, ultimately impossible challenge by a series of virtuoso performances. On the lowest level it reflected pride of craft, a desire to demonstrate professional competence. At the same time it incorporated an acceptance of war as the servant of state policy rather than as an end in itself. It is ironic that precisely the German military should be so sharply criticized for developing an ethic of total war, without corresponding recognition of the limitation of ends that for over a century accompanied the steady escalation of means. This intellectual framework was the product of interest as well as principle. Prussia's and Germany's military leaders were well able to calculate their resource position compared to their neighbors'. The principal result of that calculation was an institutionalized urge to force the pace of the game, and thereby determine its nature. But to be effective this initiative had to be sustained within a framework ultimately determined by the diplomats and statesmen. The absence of this framework, rather than the nature of the Nazi regime, posed the ultimate challenge to Germany's military establishment in World War II.

The Wehrmacht produced an ample share of brilliant craftsmen— Manstein, Dönitz, Wolfram von Richthofen. It developed no senior

officers with the vision or the will to transcend the limits imposed on
Germany's war effort by geography, politics, and diplomacy. Arguably
since the winter of 1941, certainly after the disaster of Stalingrad,
their operational focus increasingly became an escape hatch for men
recognizing the collapse of what passed for Nazi grand strategy, and
reasonably aware of the probable consequences of that collapse.[73]
Hitler's marshals responded like short-money players in a table-
stakes poker game, seeking operational victories to avert the end as
long as possible while demonstrating their own skills and character.
None passed the test of their profession at its highest levels: the test of
moral courage. By the long-established principles of the officer corps,
an unacceptable challenge to one's integrity or judgment generated a
corresponding duty to refuse positive compliance no matter the consequences. Whatever their individual motivations, the Wehrmacht
leaders remained a step below the requirements of their profession,
morally as well as technically. It is too extreme, too unacademically
idealistic, to suggest that limitations in one area reinforced as well as
reflected shortcomings in another?

Arms and Alliances: French Grand Strategy and Policy in 1914 and 1940

Douglas Porch

The task of explaining the erratic course of French policy and the fate of French arms in the twentieth century is not always an easy one. The first difficulty springs from the fact that one must confront a wall of "Anglo-Saxon" skepticism and prejudice. The French are perceived both as difficult allies, and as soldiers whose martial abilities have been rather on the decline since Waterloo. The second problem occurs when one is called upon to explain the dramatically different outcomes for France of the two world wars. How can a country which defeated the Schlieffen plan on the Marne in September 1914 and went on to fight with such stamina and courage for four years, have collapsed so utterly in May and June 1940? So great were the apparent differences in political courage, popular morale, and military performance between the two epochs that it may well be wondered if we are dealing with the same country.

Indeed, a traditional explanation for the enormous disparity in military performance assumes that the fundamental outlook of Frenchmen altered substantially between 1914 and 1940, that the men of 1940 were simply not hewn from the same timber as their fathers of 1914. Those who marched out (or who took taxis) apparently with an enthusiasm bordering upon delirium to meet the Hun in 1914, were sons of the era of high patriotism, hardened to firm resolve by a generation's worth of intense nationalist propaganda, backed by a political system which at the very least recognized the dangers for France of German hegemony. Alas, before the ink was dry on the Versailles agreement, the French had changed. The year 1914, and the "nationalist revival" which had preceded it, seemed to be merely a spasm of chauvinism which, once passed, again found Frenchmen squabbling among themselves, or prepared to relapse into a single-minded pursuit of *les plaisirs de table* and other carnal delights, until shaken from their self-indulgence by the next German offensive with its inevitable consequences. Then, after the 1940 defeat, as after that of 1870-1871, they passed through the usual rituals of contrition under the paternalistic guidance of a venerable marshal of France, which consisted essentially of much church-going and outdoor calisthenics.

This generational explanation of France's poor, not to say disastrous, performance in 1940 is, on the surface at least, a plausible one. After all, France survived the Schlieffen plan in 1914, but in 1940 she collapsed with a suddenness which, even with a half-century's worth of hindsight, still seems shocking. However, according to the Canadian historian J. C. Cairns, this generational explanation has given rise to at least three basic perceptions about the 1940 defeat which have proved resilient despite the best efforts of modern scholarship to test them: first, the "legend" of a particularly rotten regime; second, that of a particularly rotten army; and third, that of an unnecessary surrender.[1] The publication soon after the war of the French historian Marc Bloch's *L' Etrange Défaite* (Strange defeat), a powerful "J'accuse" of the late Third Republic written by a man of the left who perished in the occupation, gave to the generational argument an extra dimension of credibility. In it, Bloch blasted the intrigues and poor planning of the politicians, the pessimism and lack of faith in the country which gripped the middle classes, the low productivity, selfishness, and lack of patriotism of the trade unions, and the mental ossification of a geriatric high command and an officer corps which failed to "keep their minds supple enough to retain the power of criticizing their own prejudices."[2]

The Third Republic does have its defenders, who argue that, with a people who exhibit such demonstrably centrifugal tendencies as the French, the regime struck just the right balance of utter chaos and mere confusion.[3] The assumption which has anchored much of the historical literature on twentieth-century France has been that the Third Republic of the *après-guerre* was on the brink of disintegration, while the regime of 1914 was merely in an arrested stage of debilitation. This view is based in part on the belief that the men who led France in the interwar years were simply not of the same caliber as those of the earlier political generation—Léon Blum was a pale and inconsistent imitation of Jean Jaurès, Edouard Daladier was hardly a "Père la Victoire" in the manner of Georges Clemenceau, nor, when the crunch came, could Paul Reynaud offer the bedrock stability of President Raymond Poincaré. Furthermore, these men compounded their personal failings by enfolding the complex economic, social, and political issues of the interwar world in the outmoded perspectives and threadbare language of the pre-1914 era. As a result, the leaders of Third Republic in the 1930s presided over a deeply divided society, one which seemed to lurch between the prospective of a fascist takeover, like that allegedly attempted in 1934, and outright civil war. No surprise, then, that France failed to resist the German onslaught in 1940.

The Third Republic in the 1930s certainly had its problems. However, were the differences in the regime between 1914 and 1940 crucial? Had the Third Republic declined so far beyond the point of redemption that the German attack, while tragic in its consequences for the French people, offered almost a merciful release from a regime which was already moribund? For, after all, the Third Republic had become a music hall joke well before the turn of the century, plagued by the same problems of ministerial instability and political mediocrity which characterized the interwar period. One need only think of the Viviani cabinet of 1914 and of the fact that, throughout the July crisis of 1914, the French government was more concerned with the Caillaux scandal than with the threat of European war, to realize that French political instability did not originate at Versailles in 1919. The Dreyfus affair had deeply divided French society. And although the nationalist revival of 1911 may have injected much-needed resolve into the French body politic, historians are divided on its influence and effects. Its most noted achievement—the three-year military service law of 1913—became a central issue in the July 1914 elections, and, had the war not broken out, the left-wing victory in those elections would almost certainly have insured its repeal. In fact, so unsure was the French government of popular support for the war that it had drawn up the Carnet B, a list of Socialist and labor leaders to be arrested on the outbreak of war, lest they sabotage mobilization.

When this situation is contrasted with that of 1940, the conclusion can be drawn that there were more similarities than differences. The disasters of 1940 give the impression that the regime was on its last legs. However, this view must at least be modified. Daladier's almost two-year tenure as prime minister from April 1938 gave the government the elusive stability it had so long sought in vain. He was popular, and contemporaries believed that he had put the regime on an even keel by 1939. The extreme right was split and on the decline, while the parties of the left appeared to be in serious electoral difficulties. This, of course, did not make the regime strong—the divisions of 1936 had left too much rancor for that. But Daladier did leave many contemporaries with the impression that government responsibility had at last returned.[4]

Furthermore, the "generational decline" hypothesis assumes that political degeneration led to factional politics in the 1930s that precluded actual attention to adequate military investments. However, it is difficult to cite any specific acts which contributed to the defeat of 1940. The Communists were, of course, a new political factor. However, between 1934 and the Nazi-Soviet pact of August 1939 they generally supported French resistance to fascism. The Socialists ritualistically voted against the defense budget, yet historians of the Popular Front have successfully refuted claims that Blum was soft in defense; the French from 1936 began to invest heavily in armaments production so that, by 1940, according to the French historian Robert Frankenstein, their output attained "quite spectacular summits."[5] France spent a greater percentage of her GNP on defense than any other country between 1919 and 1935, with the possible exception of the USSR. Military spending in 1938 was in real terms 2.6 times what it had been in 1913.[6] Therefore, the Third Republic in 1940 was perhaps not the defeatist regime being led away meekly to its abattoir, as popular imagery, and perhaps some historical literature, would suggest.

The second of Cairn's "legends" is that of the rotten army. The unfavorable reputation of the French army comes largely out of this 1940 defeat. So stunning was it that, when stripped of the grosser charges (like treason), the French army merely stands accused of muddle and incompetence on a Himalayan scale, of ignoring technological innovation, of a "Maginot mentality" which demonstrated that in terms of operational doctrine the high command had never abandoned the trenches of World War I. Strategically, their advance into Belgium and Holland laid them open for a German riposte through the Ardennes which they had dismissed as impossible.

Two things might be said about this view. The first is that, if the French army received a sound thrashing in 1940, it was certainly not because they ignored new weaponry or attempted to reproduce the battlefields of the 1914-1918 war in conditions which were entirely inappropriate. The American historian Robert Doughty has recently offered a picture of an army which had struggled manfully to come to terms with the technical innovations of the interwar years. The trouble was that they sought to fit these innovations into their operational doctrine, which called for a tightly controlled battle and allowed little room for flexibility and surprise.[7] Furthermore, far from being a weak character as he is often depicted, new research has shown French commander in chief Maurice Gamelin as firmly set upon guiding the French forces toward independent tank divisions despite the maze of inter-arm rivalry and left-wing hostility to any tactical, operational, or strategic doctrine which hinted that the high command might be planning a repeat of the bloodletting offensives of World War I. He objected to the demand by Charles de Gaulle and Paul Reynaud that France create mechanized divisions, not because he opposed mechanization, but because those two men linked the issue of modernization to that of the professionalization of the army, a prospect calculated to throw the left into fits of hysteria.[8]

While Cairns is no doubt correct to dismiss the charge of "rottenness" laid against the French army of 1940, it is undeniable that the French military was beset by problems in the 1930s, and that they were serious ones. But was the French army any more efficient in 1914? Historians of the pre-1914 army are divided between the pessimists and those who seek cautiously to rehabilitate the army of those years. Although it must be confessed that much fruitful research remains to be done on French operations and tactics before and during World War I, for the moment the verdict on French performance must remain rather bleak. Tactically, the army's powers were atrophied by poor training and poor leadership. Operationally, the army proved extremely poor in inter-arm cooperation and control of battle, which meant that in 1914 the fight often deteriorated into "isolated brawls" in which the French took heavy casualties. Plan XVII, devised by French commander in chief Joseph Joffre and his staff just a few years before the outbreak of the war itself, was strategically even less realistic than that of 1940, in that it was based on an utterly erroneous view of French capabilities and of German intentions. There has, of late, been an attempt to cast the French offensive of 1914 in a more favorable light. Some in France appear to have seen offensive action as the best means of insuring early Russian intervention.[9] In military terms, the offensive sought to deny the Germans the strategic initiative and, in its orientation, allowed Joffre to recover from his initial mistake and redeploy, an option which the Dyle/Breda plan of 1940 denied to Maurice Gamelin. The trouble with the French, this argument goes, was not their offensive-mindedness but their inefficiency.[10]

The first point to make is that, although the French General Staff obviously hoped for early Russian offensive action, in December 1911 they believed that they must hold off a German onslaught alone for at least a month.[11] Joffre found the situation little improved when he visited St. Petersburg in August 1913. Although he was well received, "we felt even in the entourage of the czar a group who probably gave us external expressions of friendship, but who regretted seeing the leaders of Russia so obviously oriented toward France. The war minister, [V. S.] Sukhomlinov, in particular, promised everything but delivered nothing." Joffre also found that "the situation was delicate" for the Russian army. Their maneuvers "seemed to us organized for parade effects, without bothering to take sufficiently into account the realities of war." And while the frontier corps could be concentrated by the fifteenth day of mobilization, they would be extremely vulnerable to a German attack until the twenty-third day, when twenty-four to twenty-seven corps could be concentrated. The reserve divisions could come on line only on the twenty-sixth day.[12] Furthermore, St. Petersburg balked at French insistence that a 1913 railway loan be tied to a Russian commitment to fight Germany. So although French and Russian experts might settle on a plan, no action followed.[13]

The problems of getting a firm commitment out of the Russians were no mere illusion, for the Russian government did not speak with a single voice. Sukhomlinov jealously guarded what he saw as the prerogatives of his ministry, and refused to communicate to the minister of foreign affairs and to the prime minister the protocols of French-Russian General Staff talks of 1912, or to enlighten them on the true state of the Russian army. In foreign capitals, Russian ambassadors and military attachés sent conflicting reports which spread confusion at the highest levels of government and further increased tensions between the Foreign Office and the War Ministry. Upon this government which was often working at cross-purposes, the weak and vacillating Nicholas II was unable to impose order. Likewise, the Russian General Staff was seriously divided on whether to concentrate on Austria-Hungary or Germany in the event of war—as late as 13 July 1914 the Russian chief of staff Zhilinskii indicated to Joffre that he believed Austria to be the more menacing enemy because "the moral effect" of defeat at the hands of Vienna "would be disastrous." Even the assistant chief of staff, General Danilov, who conceded that Germany posed the greater threat to Russia, argued that Russia should carry out a strategic retreat in the face of a German attack, a strategy which would not have helped France one bit. As for the Franco-Russian staff talks which were held annually from 1910, the American historian William Fuller has written that "it would be hard to conceive of a greater degree of mendacious game-playing," with each group attempting to divulge as little as possible about their war plans. Indeed, the Russians were deeply suspicious of the French, whom they saw as trying to prod Russia into actions which served French, rather than Russian, interests, and this distrust extended even to the intelligence supplied to the Russians by Paris. Those actions included going to war in the first place to save France from a German invasion, with a military machine which the Russian generals realized was acutely defective, and against Germany when many perceived their primary threat as coming from Austria.[14]

Given the relative uncertainty about the actions of her allies on the outbreak of war, the French should have adopted a strategy which sought to preserve their army. This was especially true as the French knew that they would be the object of the initial German attack. So although a Russian offensive may have made sense on a strategic level because it would help to relieve pressure upon France, France should have sought to husband her forces until Russia could intervene. Perhaps a French offensive in 1914 might have been logical had it moved into Belgium, simply to take the battle away from France; but this option had been vetoed by the government, fearful of British disapproval. However, to fling a French army unsupported by heavy artillery against prepared German positions in Alsace and Lorraine could only have produced one result—utter failure. Moreover, it took the French further away from the main axis of German advance, which Joffre knew would be through Belgium. Last, the inefficiency of the French army should have caused its leaders to be more cautious, to be reluctant to undertake a strategy which they were tactically and operationally unable to carry out. In retrospect, the victory on the Marne may indeed be termed a "miracle," due more to German mistakes than to French efficiency. After all, a strategic plan which resulted in a 25 percent casualty rate in the first six weeks of the war and left the enemy firmly implanted deep in French territory, in possession of a great deal of heavy industry, can hardly be termed a success.

The last of the "legends" concerning the French collapse of 1940 is that of an unnecessary surrender. This refers to the alleged absence of morale in France in 1940, both popular and political, in contrast to the *"On les aura!"* spirit of 1914. The memories of the catastrophe of World War I certainly meant that French enthusiasm for conflict had cooled in the interwar years. But then, war was unpopular *everywhere* in 1939, including Berlin. Although the Munich agreement of September 1938 was greeted with some relief, by the summer of 1939 public opinion in France no longer supported appeasement.[15] The length and inactivity of the *Sitzkrieg* may have dampened spirits somewhat, and some French troops did indeed run away in 1940. However, outside a few fringe groups, there is no evidence that French opinion was defeatist *before* the overwhelming success of the German offensives became known.

And again, if placed in perspective, French popular morale in 1940 may not have differed greatly from that a generation earlier. The French historian Jean-Jacques Becker has disputed the received view that French morale was rock solid in 1914. He labels the notion that the nationalist revival revolutionized public opinion before the war as "utterly disputable." The French marched off to war with "indignation" at the German attack rather than with enthusiasm. French morale hit its nadir at the end of August 1914, as the badly mauled French armies fell back toward the Marne, the government scuttled to Bordeaux, and the Germans advanced toward Paris. So desperate was the situation believed to be that, according to Becker, popular opinion "was ready to abandon itself to the idea of an inevitable defeat." French morale was "extremely fragile . . . reacting a little like a weather vane to the gusts of wind." The major difference between 1914 and 1940, Becker argues, is that in that earlier year Frenchmen believed that the war would be short, while in 1940 they could only look forward to a long and arduous campaign.[16] Some French troops ran away in 1914—the problem of windy behavior was especially acute in the Fifteenth Corps—but most troops in 1914, as in 1940, fought with great bravery.

Should France have fought on from North Africa in 1940? With the great advantage of hindsight, my answer would be yes. Most of the high command certainly opposed the continued struggle. Both soldiers and some politicians seemed motivated by the desire to maintain order in France, to keep the empire intact, to settle political scores with their enemies on the left, and by the not implausible conviction that their only ally—Great Britain—was finished. If ultimately this proved a disastrous option, who could deny that, in the confused conditions of June 1940, the decision was a defensible, if hasty, one?

The point I am making is that the difference in military performance between 1914 and 1940 cannot be explained by cataclysmic changes in the political structure of France, in the quality of the French army, or in popular morale. For, if the outcomes were different, there seems to be a great deal of consistency in the fact that the strategic and operational choices made by the French army in 1914 as well as 1940 were fairly disastrous. If one accepts the notion that France survived perhaps only by the narrowest of margins in 1914, then the real ques-

tions should be: How can we account for the relatively poor performance of French arms in the twentieth century? Is it true that French
generals always fight the last war? Or were there other factors which
distorted their strategic vision and blurred their judgment?

Perhaps the best way to deal with such issues is to inquire whether
there are threads of continuity in French strategy in this century
which help to account for this uneven record. In other words, by
studying some of the components of strategy—alliance systems, civil-
military relations, and operational doctrine—can one explain the in-
adequacies of French performance?

Of course, one way to answer the question is simply to turn it
around and ask, Why did the Germans do so well? For, when placed in
perspective, victories over Germans in the opening stages of the two
major wars of this century were the great exception, and the French
may claim one of them. Defeats of the Allied powers were the rule. No
one did well against Germany in the opening stages of these wars, and
that includes the U.S. Army and the Red Army. Two conditions of
German success help to account for her initial victories. First, Germany was strong enough to be able to take her strategic decisions
without having to consult powerful allies. It will escape the attention
of no one that this was a severe long-term weakness. It might also be a
short-term weakness, as, for instance, in 1914, when a coordinated
plan with Austria-Hungary to attack Russia might have saved Germany some desperate improvisation, improvisation which some
might argue actually cost her the victory on the Marne. And in 1941
Operation Barbarossa might have enjoyed more success had Japan
gone to work in Siberia. But at least German power was sufficient in
the short term to mean that she was not hamstrung in military operations by the need to coordinate strategies with reluctant allies.

A second factor in German military success was the ability of her
forces to develop an effective operational and tactical system. Again,
it could be argued that this was a long-term weakness, for operational
and tactical efficiency did duty for strategic thought in the German
army. However, a system which combined effective inter-arm cooperation with the devolution of responsibility to lower levels of command gave the German army great flexibility, especially in "encounter" battles where individual initiative and judgment were so
important.

Why was the German army able to accomplish this? There are many
reasons—historical, a strong operational tradition, perhaps even the
intervention of Hitler. Essentially, the German army operated in a
political environment which not only encouraged but *demanded*

maximum military efficiency. The fact that the army historically had been "the pillar of the Prussian monarchy," that German unification had been brought about through war, and that Germany's precarious geographical position in the center of Europe required her to maintain a strong defense made the army largely immune from serious criticism before 1914. After 1933, of course, the army again became the focus of official favor.

With this in mind, we can return to the question of why French military performance was so uneven in the first half of this century. The evidence suggests that the origins of French deficiencies can be traced to two factors which influenced French strategic planning and the French army's operational and tactical efficiency. First, France has been a country in relative decline in this century. Second, French soldiers have had to take into account that they served a politically divided society.

That France was a country in relative decline had been obvious to the more thoughtful even before the Franco-Prussian War shattered any illusion left over from the wars of the French Revolution and empire about the innate military genius of the French people. The beneficial aspect of German unification was that it allowed France to identify her major enemy. The bad news was that Germany was growing steadily more powerful in manpower and industrial resources, while the French seemed reluctant to procreate and industrialize with the same vigor as their neighbors. France's problem became that of matching ends to means. Before 1914 she was forced to spend a greater percentage of her budget on defense and conscript a larger percentage of her manpower than Germany. Still, the shortfall was significant. De Gaulle lamented in the 1930s that two Germans were coming of military age for every Frenchman. How could France offset the German advantage?

One way was to conscript manpower in France's colonies. In 1910, Colonel Charles Mangin argued in his book *La Force noire* that Africa offered a vast reserve to fill France's manpower deficiencies. However, the French were slow to take advantage of this resource for several reasons. First, colonialism was not popular in France. Second, settlers, especially in Algeria, opposed it, and colonial administrators feared that native conscription would provoke rebellion. Nor was this solution to the manpower problem popular in the army, where many French generals considered colored troops inferior. Last, the left resisted the importation of mercenary troops into France, fearing that they would be used as strike-breakers. Only in 1917, with the arrival of Clemenceau as prime minister, did France begin to tap her colonial manpower reserves on a large scale.[17]

However, colonial manpower could offer only a minor palliative, even in 1940 when France made a better job of mobilizing her colonial subjects. The chief potential source of extranational strength for France lay with her alliances. Before 1914, the French Foreign Ministry spent enormous energy trying to break out of the diplomatic isolation which Bismarck had imposed upon it. Those efforts brought success with the Franco-Russian alliance of 1894, followed by the Entente Cordiale of 1904. Italy was also neutralized as a potential adversary from 1902, while agreements with Spain over Morocco removed the major source of diplomatic friction between France and her neighbor to the south. After 1919, that tradition reasserted itself. Indeed, in the interwar years the search for alliances became almost obsessive. But how important were these alliances to military planning? It is worth arguing that French defense planners paid too little attention to alliances before 1914, while in the interwar years defense planning became virtually the hostage of the demands, real or imagined, of alliance politics. This is perhaps an ironic conclusion, given that France's allies virtually assured her survival in 1914–1918, but when the crunch came in 1940, she remained alone and defeated.

In 1914, French soldiers failed to make alliances the foundation of their military planning for several reasons. First, there was little coordination between the Foreign Ministry and the War Ministry. For instance, the November 1902 accord between France and Italy which helped to separate Italy from the Triple Alliance was not communicated to the War Ministry until June 1909, so that the French continued to maintain a substantial army on the Alps.[18] Until 1911, the Quai d'Orsay remained largely ignorant of the Franco-British staff talks begun in 1906. Only in the wake of the Agadir crisis was a Conseil Supérieure de la Defense Nationale established to provide a forum for politicians and military men to exchange information. But the coordination of military and foreign policy was never completely satisfactory, nor the distrust between the departments completely erased, as the confusion during the July crisis of 1914 was to prove.[19]

Second, attempts to initiate joint planning with the British and especially the Russian high commands often did not get very far due to secretiveness, or lack of trust, or were not taken seriously because, although they might facilitate military intervention, they would not guarantee it.[20] Last, French soldiers did not rate the military capabilities of their allies very highly. In the opinion of Christopher Andrew, "France's conviction that she would win the war thus depended less on the assistance expected from her partners in the Triple Entente than on the belief that, in the decisive early weeks, she could, if necessary alone, secure a decisive advantage over Germany on the western front. . . . France's growing confidence in victory thus reflected chiefly a growing confidence in herself."[21]

This is not to say that alliance politics played no role in French strategic planning in 1914. Samuel Williamson has demonstrated that, although the prospect of British intervention was not responsible for Plan XVII, it did influence the direction of the French offensive by forcing Joffre to point his attacks toward Alsace and Lorraine rather than into Belgium, where a premature violation of neutrality might keep Britain on the sidelines.[22] Gerd Krumeich has written that Poincaré believed that an offensive strategy offered the best guarantee of Russian entry, although that interpretation goes too far in its insistence that French war planning was "dominated" by the needs of the Russian alliance.[23] On the contrary, Joffre seems to have employed the critical clause of the 1892 Franco-Russian convention (which obliged the two allies "to thoroughly engage [their respective forces] with all speed" as a way to justify and reinforce his previous decision to take the offensive upon the outbreak of war.[24] Although French soldiers clearly hoped for both British and Russian assistance, they were under no illusions that they must bear the brunt of the German offensives in the opening weeks.

The lesson taken away from World War I was that France's survival as a nation depended upon the strength of her alliances. Alas, this lesson was applied in conditions which were less favorable than before. It is tempting to conclude that France played her alliance cards much better before 1914 than before 1940. And in many respects she did. But it is only fair to note that the hand she had been dealt after 1918 was a very weak one. Despite her defeat, Versailles had left Germany the war's strategic victor. Her industry and population were intact. Whereas in 1914 she had been surrounded by powerful neighbors, now only France of the major powers shared a border with her. A jigsaw of small states occupied her eastern and southern frontiers, while Britain demonstrated an extreme reluctance to provide for her own defense, much less intervene on the Continent. The French have been accused of placing ideology over survival by showing little enthusiasm for resurrecting their alliance with Russia in the 1930s and by demonstrating a lack of resolve in the 1936 Rhineland crisis, although it is now clear that the Soviet leadership had no intention of fighting Germany.[25] Italy was mercurial and ideologically had more in common with Hitler than with the Third Republic.

This was hardly a propitious international climate in which to construct a system of defense alliances, and France never resolved her dilemmas, perhaps because, short of taking unilateral action which had been discredited by her Ruhr occupation of 1923, those dilemmas were insoluble. An independent foreign policy would have forfeited any hope of British support, as well as run the great risk of condemnation by those Frenchmen who feared another war. The difficulty for the French in coming to the aid of their Eastern European allies was brought home forcefully during the Czech crisis of September 1938. The Petite Entente foundered upon the traditional animosities of the peoples of Eastern Europe, and the refusal of the Poles and Rumanians to allow Soviet forces access to their territory, even had the Soviets been inclined to honor their 1935 commitments.[26] Furthermore, the French were reluctant to adopt an offensive military strategy which would have forced Germany to fight on two fronts. Given the unfortunate experience of Plan XVII and the subsequent attacks through the Nivelle offensive of 1917, the civilian leadership would certainly have balked at any suggestion that history might be in the process of repeating itself. In this respect, the defensive was a national preference. It was also a military one: the feeling among French generals after 1935 that the military balance was tilting against them, that the German army was growing progressively stronger behind the barrier of a fortified Rhineland, and that French troops were poorly trained and therefore bound to be at a disadvantage in offensive action, increased their reluctance to act. In a Clausewitzian sense, the defensive was logical because it was the option of the weak. At the same time, doctrine was recast to emphasize the traditional French notions of controlled, methodical battle and the superiority of firepower over maneuver, the very concepts which had almost forfeited victory during the Michael offensives of 1918. [27]

Therefore, even after Great Britain was enticed into giving a guarantee to Poland, the Allies remained glued in place. If one accepts that an Allied offensive into the Rhineland in September 1939 would have done more damage to the attackers than to the defenders, and would not have prevented the collapse of Poland in the bargain, then perhaps the decision to sit tight in the Maginot line was defensible, especially had Hitler unleashed his armies toward the west in the autumn as he had originally planned. Williamson Murray has made the case that Allied strategy, while sound in conception, was poorly executed. The Allies correctly identified Germany's economy as her Achilles' heel, yet they failed to act decisively to cut off her outside conduits of supply and simultaneously to force her to fight, thereby using up her scarce resources.[28] Although such a view seems plausible with all the benefits of hindsight, to the men at the time the implications of forcing Italy into belligerence when they already had their hands full must have appeared daunting. The historical premise upon which a call for more forceful Allied action rests is that the Germans were getting stronger all the time, that any procrastination could, and did, only benefit Hitler. Yet, would this have been the case had the German offensive of May and June 1940 been contained in Belgium or even in northern France, as the Allied leaders had a reasonable belief that it would and could be? How long would Germany then have been able to hold out? The Dyle/Breda plan was determined in great part by Gamelin's desire to assure the assistance of Belgian and Dutch troops, and to move the battlefield as far as possible away from France's richest economic areas, something which had not been done in 1914. This strategy failed because, ultimately, the Allied armies did not have the operational and tactical ability to contain a surprise German thrust through the Ardennes.

Why the French forces should have been so badly prepared to counter those of Germany relates in great part to the second difficulty which confronted French soldiers—that they must prepare defense plans in a politically divided society. Even more than alliance politics, the political debate in France over the place of the army in the state influenced the French operational and strategic approach to war. Of course, debates over military policy are common in many countries and can even be healthy if they challenge preconceived notions and dissipate "group think." However, in France the acrimonious debates over the role of the army in the nation made it difficult to establish a defense policy based upon a clear analysis of strategic realities. Everyone agreed on the need for a defense policy. But political divisions distorted the debate.

Traditionally, the left favored a broad-based army of short-service conscripts. In their view, such an army offered the best protection against the foreign adventures and coups d'état of right-wing militarists. If the nation were attacked, then Frenchmen would rise up as one man to throw back the invaders, just as the armies of the French Revolution had done. The right, on the other hand, argued that a small professional army was both more efficient and the best guarantee against social disorder. The Dreyfus affair crystallized the debate over just how autonomous, just how free from political interference, the army should be. A book by the Socialist leader Jean Jaurès, *L'Armée nouvelle*, written in 1910, broadened the debate to include strategy. This associated the left with a defensive strategy which, he argued, was the best suited to an army of poorly trained conscripts, and ran fewer risks of escalating international tensions. The costs of World War I seemed to validate Jaurès's view that professional soldiers needed strict control lest *esprit militaire* be allowed to run amok, and generals sacrifice troops to no purpose other than for their own glory and promotion. In the 1930s, therefore, the offensive found few supporters in political circles, where the defensive was considered less costly in manpower, more suited to a democratic nation, and less of a threat to the stability of the international environment. This doctrinal hesitation to act was reinforced by the belief, already alluded to, that unilateral international action by the French would forfeit any hope of British support, although how this squared with France's eastern alliances was a dilemma which was never resolved, especially after Italy moved closer to Berlin.

The question which remains to be answered is, how did service chiefs react to this increasingly hostile political environment? By applying concepts of organizational theory, Jack Snyder in his study of the prewar European armies has argued that the doctrine of the offensive caught fire in the pre-1914 French army because it best served the interests and self-image of the French officer corps, whose prestige and morale had been savaged by the Dreyfus affair. It was also a budget ploy because it was more expensive than the defensive, and it allowed military leaders to simplify complex data and to reduce the uncertainty of battle.[29] In his study of the interwar years, Barry Posen, while relying more on balance of power "theory" than does Snyder, makes essentially the same arguments. The popularity of defensive doctrines in the 1930s, he writes, helped to protect the vulnerable political position of the officer corps because it was what the civilians wanted. Therefore, it assured the autonomy of the army. The defensive had a financial component because "buck passing" is a consistent feature of alliance politics. Last, the defensive reduced uncertainty by allowing officers to refight the sort of war they had learned in 1914–1918. It also minimized the lack of training of French short-service conscripts by putting them in bunkers.[30]

This political science approach certainly allows us to pick out strands of consistency in French strategy. The traditional view as explained by the "nation-in-arms" historians assumes that the French army was a sort of aristocratic enclave keen to perpetuate a reactionary institution in a democratic world. Organizational theorists argue that organizations attempt to perpetuate what they perceive to be their own self-interests. But do these views stand up to close scrutiny?

The first argument states that the strategies adopted are a bid for the autonomy of the professional officer corps. Of course, what "autonomy" means is not clear. One might concede that the French army sought "autonomy" insofar as it wished to be exempted from the political feuds of right and left. But this does not mean that French officers wished to live as some sort of aristocratic fraternity in a lower middle-class republic, or even as a bureaucratic fortress, as some social scientists believe.

This is no army which desired splendid isolation. On the contrary, it yearned for acceptance. It desired nothing less than spiritual union with the French nation, to become the focus of French patriotism and pride, the institution in which dwelled the hopes and aspirations of all Frenchmen, of whatever political persuasion. Consequently, the strategies of two world wars were not strategies of autonomy, but of integration. In 1914, the offensive was the strategy which divided Frenchmen the least. For conservatives, the doctrine best suited their military traditions. For the left, especially the radical left, it revived the "moral force" based upon patriotic enthusiasm which had confounded the enemies of the French Revolution. It also satisfied the increasingly strident calls of colonial soldiers for a doctrine which would announce the revivification of the French army after the demoralizing day of the Dreyfus affair.[31] As for the interwar years, it would seem to require some circuitous logic to see the defensive as a bid for autonomy. Weygand certainly worried that the professional army might suffer great depredations at the hands of the left.[32] But, all the more reason to see the defensive posture of France as one of political integration. And, as has been noted, De Gaulle's theories were opposed by Gamelin, not on military grounds, but on the political ones that they jeopardized the very integration they were striving so hard not to upset.

The second argument contends that strategies are merely budget ploys devised with financial, rather than strategic, goals in mind. However, the relationship between the two has not been conclusively proven. Before 1914, the offensive was the cheap option. In 1913, the High Command was forced to postpone requests for heavy artillery and fortress construction to get the three-year service bill which Krumeich sees as the sine qua non of Plan XVII.[33] This virtually forced the French to adopt a strategy of mobility because they were armed only with rifles and light artillery. A defensive strategy based upon fortress construction and heavy artillery would surely have been a bigger money spinner.

The question of financing French forces in the interwar years remains mired in controversy. Posen seems to suggest that the defensive was the cheap option because it allowed France to pass on the costs to her allies. The notion of buck passing as a feature of alliance politics seems questionable. Leadership in an alliance usually falls to the partner willing to bear the greatest burden. Also, strong nations make more attractive partners than do weak ones, which is one reason why French generals paid so little serious attention to their allies before 1914. Nor did alliance politics prevent France from spending a great deal on defense. But again, if strategies are merely budget ploys, then surely it would have paid the French High Command to develop their own blitzkrieg doctrines to gouge money out of parliament for tanks and planes.

The last argument, that the strategies chosen simplified data and removed uncertainties, offers perhaps the best argument as far as it goes. The offensive doctrine of 1914 allowed the French High Command to ignore weaknesses in their own forces, to end the tedious debate over tactics which the army seemed unable to resolve, and to assume that French soldiers could compensate for poor training with "moral force." The more defensive strategy of the interwar years sought to minimize the uncertainties of "encounter" battles with poorly trained soldiers and the perceived inferiority in tanks. But if taken too far, this argument assumes that military culture is an impediment to innovation—esprit militaire, aggressiveness for its own sake, the scramble for promotion, the tendency to fall back upon tested and true values when confronted with complex situations brought about by technological change. This view holds that soldiers can be brought to drink from the troughs of innovation only when forced to do so by civilians or confronted by defeat.

No doubt this is sometimes true. But there are many examples of military organizations innovating on their own accord, as did the Wehrmacht between the defeat of Poland in 1939 and the invasion of France in 1940, even though it had been successful in Poland. According to the British historian Martin Alexander, Gamelin probably did more to push the French army toward reform than did any civilian. Of course, failure to innovate is the bureaucratic norm. When placed in context, the French can be said to have done no worse than anyone else against Germany in 1914 and 1940. The army's inability to do better appears to be less the reflex of a particularly stunted professional milieu to the technical, administrative, or budget questions which confronted it, than the confused response of an army operating in an often openly hostile political environment. Whereas the Germans made the political choices necessary for military efficiency in a clear-cut manner, the French response was more equivocal. The result was muddle and indecision.

Civil-military friction deprived the military leadership of authority, both before 1914 and before 1940. The High Command was an amorphous group over which the commander in chief presided but did not rule, where authority was allowed to wander in a bureaucratic maze for fear that a concentration of military power in too few hands might bring on unwanted health problems for the Third Republic. This was the direct result of the fear of an overly powerful military influence in politics. Therefore, debates over strategies, operational concepts, and technological innovations had to involve compromise, for no one was able to establish clear priorities. One bureau's gain was another's loss. This caused problems both in adopting heavy artillery before 1914 and in establishing tank divisions before 1940.

The political climate was often unfavorable to military innovation. The request for funds for heavy artillery was postponed in 1913 because it was felt that, in the wake of the acrimonious debates over the three-year service law, the time was not propitious.[34] De Gaulle's clarion call for a professional mechanized army jeopardized the patient attempts of Gamelin to prod the army, and parliament, toward a more modern tank doctrine. Ministerial instability further weakened the forces' ability to reach decisions and establish priorities.

The operational and tactical expertise of the army was seldom good enough to achieve their strategic goals. Plan XVII perhaps offers the best example of a tactical/strategic mismatch. There can be no question that French generals in both wars made appalling mistakes. The question is, how far was political prejudice responsible for this? For instance, it has been charged that Joffre's failure to use reservists more fully in 1914 sprang from a professional commitment to the long-service regular army.[35] It proved to be rather like going to war without your trousers on, for the French were outmanned and outflanked by a German army that threw its reservists into the front ranks from day one. But Joffre's hesitation to use reservists in 1914 was based on a sound belief that they were simply not battle-ready. Joffre's mistake was to "mirror-image," to fail to realize that German reserves, who were better trained and who had a far higher proportion of professional officers and noncommissioned officers, were quite capable of participating in the Schlieffen plan.

A second failing was the army's doctrine calling for a controlled battle. This demonstrated, it is alleged, an unwillingness to innovate, a desire to reproduce the familiar battles of the last war. "As I have said," declared French general André Beaufre of the French strategy of 1939–1940, "the idea was to recreate the conditions of 1918—and wait."[36] This would be going too far. However, it was equally true that the lack of training of most of the lower cadres simply did not allow the French commanders the luxury of a more decentralized battle.

One may fault them for not doing more to develop the capabilities of subordinate commanders. But this failure sprang essentially from an unwillingness to create a strong army, from the desire in the interwar years to transform the professional soldiers into a training cadre for the "nation-in-arms"; perhaps also from the feeling that things had "got out of hand" in August and September 1914, when too much initiative and too little control brought the army to the very brink of disaster. The government neglected to accord the forces the sort of prestige and material advantages which would have allowed it to maintain high morale, to train its soldiers harder, to require greater sacrifices from Frenchmen for national defense, and to work in an atmosphere of mutual trust when evolving operational and tactical doctrines, and when making strategic choices. In all of these areas the German army was superior, but then French soldiers were not competing on a level playing field.

I have argued in this essay that the poor strategic choices and inadequate operational and tactical doctrines of the French army in the twentieth century were not the result of a desire to impose a right-wing political agenda, of an outmoded military culture, or of strategies of bureaucratic self-preservation. An approach which attempts to explain the behavior of the French military leadership by assuming that they were attempting to impose some off-the-peg political philosophy, or by pillaging a few history books to prove a preset social science equation, does not get us very far. Let us assume that we are dealing with men of goodwill who were trying to do a job—namely, defend their country—but that they were forced to confront the undeniable facts that France was relatively weak and therefore needed allies, and that the political environment was often hostile. This does not excuse their strategic blunders, but it does help to place them in perspective. For, when all is said and done, French difficulties in the two wars were remarkably constant: her soldiers were required to coordinate the fight with allies whose intentions were not always clear, while the relatively inexperienced nature of her army required her to rely on a plan, and eventually a controlled battle, which lacked flexibility and failed to take the unexpected into account. In the final analysis, the differences in outcome between 1914 and 1940 are to be found more in the German ability to exploit their initial victories than in any serious decline in France's powers of resistance.

The Evolution of Soviet Grand Strategy

Condoleezza Rice

Before attempting any analysis of the various phases and features of the evolution of Soviet grand strategy, it is important to alert the reader to the fundamental difference between it and the strategies of the traditional European nations (Britain, Germany, France) that have been discussed in the preceding chapters. The difference is that Soviet strategy was from the beginning influenced by a holistic and universalist ideology, a belief system that not only viewed politics, society, economics, and warfare itself through the lenses of the class struggle, but also forecast (and strove to realize) a transcendence of existing power relationships.[1] Like a religious crusade or, perhaps, the French revolutionary movement of the early 1790s, there existed within Marxism the hope of converting the entire world into its own image; and when that transformation had taken place, the traditional instruments of power would also fade away.

This holistic belief-system brought with it both an advantage and a disadvantage. The advantage, as I will show below, is that Soviet strategic thinking was to see no great "break" between war and peace (each was a phase in the larger process of the class struggle), nor did it assume that the military dimensions of strategy could be divorced from its critically important nonmilitary dimensions. In theory at least, this gave Soviet grand strategy a continuity and a coherence all too often missing in the West. The disadvantage was that this Marxian belief in the inevitability of the world proletarian revolution some-times "de-ranged" the Soviet leadership's handling of problems and crises where the actual circumstances did not accord with the theory. Unable to reject their own ideology, the Soviets had to strive to recon-cile it with the often unpalatable facts of the existing situation.

This dilemma is nowhere more clearly seen than in the first stages of the Russian Revolution itself. Because of their ideological assump-tions, the early Bolsheviks did not think that they would have to worry about the pursuit of political goals through military power. As good Marxists, they emphasized the coming worldwide revolution following the victory of the proletariat in Russia. Workers around the world would rise up, overthrow their rulers, and construct socialism without regard to national boundaries. The international system, with its internecine wars and balance of power, were of little interest to Bolsheviks devoted to starting a chain of revolutionary events. Reg-ular armies, themselves being instruments of class oppression, should be disbanded immediately and replaced with a people's militia.

Once in power, the Bolsheviks saw that the world was more compli-cated. The workers of the world had to rise up quickly or the Germans were going to put an end to Bolshevik rule. The Bolsheviks eventually opted—temporarily, in their minds—for traditional instruments of statecraft: they quickly built a conventional army to defend them-selves and, through diplomacy at Brest-Litovsk, secured peace.

This early clash between ideological predilection and realpolitik presaged a central and continuing tension in Soviet policy. The tenets of Soviet grand strategy, or the development of a blueprint for coping with questions of war and peace, are thus unique among European great powers. Soviet leaders have had to grapple with the classical problems of statecraft with one eye riveted on their legitimizing myth—pursuit of the international victory of the proletariat.

Perhaps this legacy accounts, in part, for the absence of a Soviet Liddell Hart or Clausewitz and for their discomfort with the concept of grand strategy. The term itself is clearly alien from the Soviet point of view. Soviet strategic thinkers often disparage it and always refer to grand strategy as a Western concept. As Marshal of the Soviet Union and former Minister of Defense V. D. Sokolovskii noted, "In the West a broadened understanding of military strategy is represented as a syn-onym of politics. They have advanced the concept of 'great' or 'grand' strategy. . . . Some even extend this to all spheres of a state's foreign activity. In our country, there is military strategy and it has never exceeded the boundaries of that which is military in its broadest form . . . the theory of comprehensive preparation of the country for war."[2]

On the other hand, Soviet leaders since Lenin have been impressed with Clausewitz and his complicated explication of the relationship between war and politics. Lenin's own copy of Clausewitz is annotated, and the section on the relationship of war and politics declared "pretty clever." We know, too, that the Soviets have invested enormous resources in military power and pursue an active foreign policy drawing upon it. The Soviet Union with its great land army, led by a classically European General Staff, possesses a thoroughly traditional military instrument. Thus, as with so many things, the absence of an exact Soviet equivalent does not mean the failure to appreciate a concept. Rather, the Soviets have their own terminology that conceals their fundamental preoccupation with the classical questions of grand strategy.

Modern Soviet thought is precise in its categorization of the levels of policy that constitute "grand strategy," taken here to mean the formulation of political goals and political-military objectives, and the development of military means to carry them out. Political doctrine, which assesses the character of the international system and the Soviets' role in it, is broad and shifts in it are rare. It does provide the context, however, for military doctrine, which in turn has two sides that equate roughly with political-military objectives and military means. The political side is preoccupied first and foremost with avoiding war. But because the Soviets link the preparation of war to its avoidance, this level of doctrine also seeks to answer the questions against whom and where to fight, what resources to spend, and when to go to war. The military-technical side of doctrine is concerned with how to prepare the armed forces for war and informs military strategy or "how" to fight.

This formalized system of concepts exposes an overriding preoccupation with classical questions of grand strategy. How can war be avoided? If it cannot, what peace is desired at the end of the war? What combination of self-reliance and alliances and coalitions can be used? What resources should be devoted to the construction of military power in peacetime? Agreeing with Clausewitz, the Soviets assign these concerns to politicians. Questions of how to fight—by what means—are the prerogative of the military profession.

The effort to bring political goals, political-military objectives, and military means into harmony has been, for the reasons given above, more difficult for the Soviet Union than for any other great power. The most vexing problem for Soviet leaders has been to align two sometimes conflicting political goals: protection of the Soviet state within the system of states, and pursuit of the final victory of socialism that promises to destroy the state system.[3]

It is fashionable to say that the first has always dominated the latter and that ideology has been nothing but post hoc justification for Soviet great power interests. This proposition is too simple and belies the tensions in Soviet grand strategy. Avoidance of war and protection of the Soviet state have most certainly been the overriding concerns. Yet great powers seek not only to avoid war but also to shape the international system without resort to war. Thus, reliance on the legitimating myth of the international proletarian order has served as a constraint on Soviet leaders in their efforts to act as a great power.

There is always a danger in any post hoc recreation of how a national strategic plan evolved, and particularly so in the Soviet case, where the usual standards of documentary evidence cannot be met. Yet we know enough to reconstruct the story of how Soviet leaders have understood the world and how they have developed a strategy to pursue their interests. The strategy has had a number of important "branch points," each moving it closer to traditional notions of state interest. But it is critical to avoid, in retrospect, a sense that this was foreordained. Rather, at several points the Soviets faced key choices. Those choices, in turn, fundamentally shaped the options available to the Soviet leadership in dealing with the rest of the world.

The first key branch point may be subject to some debate. Clearly, Lenin's insistence on seeking peace with Germany over the objections of those who wished to "wage a bare-handed revolutionary war" represents the first decision that the protection of the infant Soviet state was preeminent. The Soviets had pressing security concerns, not the least of which was the growing rebellion throughout the country and the establishment of "White" governments on many corners of the collapsing czarist empire. Faced with threats on all sides, the Soviets created the Red Army of Workers and Peasants in 1918, albeit as a "temporary" measure. They were simply trying to survive, but they laid the groundwork for the creation of an army that would look much like those of bourgeois powers around the world.[4]

Although the survival of the weak Soviet state was the primary reason for the creation of the army, there were in fact flirtations with the notion that it could be the armed vanguard of the proletarian revolution. Mikhail Tukhachevsky, who ironically would become the father of a very traditional military strategy, early on suggested that the Soviet Union form a "proletarian internationalists general staff" to plan the military component of the workers' revolution.[5] Whatever fascination the Soviets held for the export of revolution by the Red

Army was tempered greatly by the failure of the "liberation" of Warsaw in 1920. Adopting a militarily risky strategy, the Soviets found that they were unable to defeat the Polish army. Their hope for a spontaneous uprising of workers was also frustrated when the Polish workers rose up, not against their class oppressors, but against the Red Army. The Red Army was lucky to stabilize a defensive line on Russian territory, and although some continued to believe that armed might would contribute to the victory of socialism, the export of revolution at "bayonet point" lost much of its luster.[6]

As the external and internal dangers to the regime passed, the fundamental question of the role of the young Soviet state surfaced again. Revolutionary tides were subsiding throughout Central Europe, and the Soviet Union found itself surrounded by hostile capitalist powers intent on isolating it from the international order, represented at the time by the League of Nations. The country, which had only begun to industrialize under the last Romanov rulers, was weakened and exhausted from the war. In these circumstances, the victory of socialism must have appeared pyrrhic at best. A great deal was now at stake because the debate about grand strategy took place in the context of a political struggle to succeed Lenin. The victory of Joseph Stalin's "socialism in one country" over the doctrine of "permanent revolution" put an end to the notion that the defense of the Soviet revolution, at the expense of armed insurrection around the world, was a temporary compromise. As such, it provided the rationale for the defense of the Soviet state as a permanent and enduring peacetime goal.[7]

Stalin believed that the revolutionary tides, which ebbed dramatically at the beginning of the twenties, were not likely to reappear soon. He argued that the Soviet Union should accept that circumstance and become as strong as possible in preparation for the next war. The only problem with the Brest-Litovsk strategy, Lenin's idea for an immediate peace with Germany at all cost, had been that the Soviet Union was too weak and suffered needlessly in its period of retreat. Stalin declared in memorable language that the Soviet Union should not be "toothless and groveling before the West again."[8] The capitalists would at some time attack, and although world revolution eventually would come and provide the final victory of socialism, a strong Soviet Union should be the young state's first consideration. Early Bolsheviks believed that what was good for proletariat internationalism was good for the Soviet Union; reversing these priorities, Stalin saw that proletariat internationalism could be made to serve the Soviet state.

Those who opposed Stalin, the proponents of permanent revolu-

tion led by Leon Trotsky, are sometimes glorified in the West. They
would seem to have had few arguments in their favor, but had they
won, the Soviet state would certainly have developed differently. Had
the goal been, for instance, promotion of revolution abroad as necessary for the survival of the USSR, the Red Army might have been
turned to that purpose. But they did not, and the Soviet Union's first
goal became to secure its borders and prepare for the next war.[9]

It is often forgotten that there was a third line of argument pursued
by the far rightists like M. P. Tomsky. They also rejected the notion of
permanent revolution but believed that Soviet goals could not be
pursued in isolation. In particular, they were early supporters of the
import of foreign capital as a central goal of Soviet policy. Thus,
although the proponents of permanent revolution disagreed with Stalin on the political goal, the far rightists disagreed with him on how
and how fast to meet those goals.[10]

When Stalin's socialism in one country triumphed in 1928, the
political goal was set and the leftists defeated. Within two years, the
means by which to meet those goals were also determined. The defeat
of the rightists meant that the Soviet Union would go it alone in the
world. It was not that the Soviet Union rejected Western help; trade
and credits played their part. But the objective of industrialization
under Stalin was to build a Soviet economic system so that economic
"warfare" could not be waged by the West. Stalin believed that dependence on capitalism was as dangerous as premature attempts to overthrow it. This meant that industrialization was achieved at very high
cost, primarily at the expense of the lives and property of the
peasantry. The brutal collectivization drive was the price that the
Soviet Union paid for rapid and independent industrialization.[11]

Stalin believed that economic strength was the first priority and
that the new industrial base could then support a massive drive to
build up the armed might of the state. With the Red Army's mission
now clearly defined as protection of the Soviet Union, military planning became a permanent fixture of Soviet life. It is not surprising, for
instance, that the first Five Year Plan was created in 1927. The Soviet
Union intended to avoid war as long as possible, but in the event of
war, Stalin was determined that Soviet military power would be second to none.

Interestingly, while committed to building up Soviet military
strength, Stalin was resistant to any reversal of priorities that would
sacrifice overall industrial power to the interests of the army. Mikhail
Tukhachevsky, who throughout his service in the Soviet army was
both admired and distrusted by Stalin, was continually rebuked for

putting the interests of the armed forces above those of the economy. Tukhachevsky was fond of noting, for instance, that diplomatic and economic plans ought to have the General Staff's approval to make certain that the goals of military power were considered. When he approached Stalin with a particularly huge request for military resources in 1927, however, he was told to revise the estimates. "The current request," Stalin is reported to have said, constituted "Red militarism."[12] In 1931, with the Japanese move into Manchuria, Stalin revised his own estimates upward and accelerated the development of the army.

Permanent institutions of military planning also came into being in this period of the early thirties. The Council for Labor and Defense, a body that had formerly been charged with overall direction of the civil war, was reorganized and subsequently renamed the Defense Commission. Political leaders like V. M. Molotov, Klimenti Voroshilov, and Stalin himself were key figures on this body, entrusted with the preparation of the country for war.

Similarly, the Red Army was reorganized and professionalized. There was at first a fierce ideological debate over whether the military forces of a socialist country should be organized into a peoples' militia or a professional army, a debate which ended with the decision to adopt a more conventional professional model. The early 1930s was also a period of rapid change on the battlefield, since Soviet strategists, like military theoreticians worldwide, were haunted by the costly trench warfare of World War I. The new technologies, with their potential for rapid advance, were thought to provide answers, but the effective use of mechanized armor was not self-evident. Soviet military thinkers eventually worked out a strategy of "deep operations" and convinced the relevant political leaders, particularly Stalin, that mechanized forces, not the cavalry, would rule the modern battlefield.[13]

As the second European war approached, the Soviet system of planning and Soviet concerns looked very traditional indeed. Political objectives included avoiding war by diplomatic means and preparing the country for war should that fail. The Soviet military, staffed by a professional cadre, was concerned with the character of modern warfare and the incorporation of new technologies. Then, abruptly, Stalin helped to destroy that which he had been determined to create. The chaos of the blood purges of the late 1930s engulfed the Soviet army as three of five marshals of the Soviet Union, and approximately 60 percent of the High Command, were liquidated. The officer corps became understandably cowed and many of its members whose lives had been preserved abandoned any attempt to give outspoken military advice.[14]

Stalin, increasingly isolated in his personal life and overimpressed with his own diplomatic skills, miscalculated badly the situation in Europe, and as a result pursued a faulty strategy. Given his own assumption—and, indeed, the basic Bolshevik premise—that all of the other great powers were antagonistic toward the Soviet Union, it was perfectly logical to "play off" the rival capitalist states and even to sign the Nazi-Soviet pact of August 1939. But given the hostility of most of the rest of the world, it was much less logical to purge the army leadership, devastate the intelligence-gathering networks, conduct a witch-hunt against all who had experience of foreign countries, and in general set back the armed forces of the Soviet state. Stalin may have achieved his political purpose of imposing ideological orthodoxy and absolute obedience to his authority, but at the cost of weakening the military instrument of Soviet grand strategy. Compounding this larger folly by his continued refusal to accept all the 1940–1941 reports that Hitler soon would attack the Soviet Union, he was in many ways lucky that the "system" survived the ferocious Nazi onslaught, albeit at appalling cost—and with the military professionals being given a belated recognition.[15]

On the other hand, Stalin's concern that even in peacetime all sectors of Soviet society and economy—industry, agriculture, the Party, the youth organizations, the press—should be prepared for a desperate fight against jealous, powerful foreign enemies did assist the overall grand strategy. In so many ways the Soviet Union's victory in World War II was a victory for his concept of a whole country mobilized for war. The civilian population, once the attitude of the Nazis toward Slavs was made clear, contributed to the efforts of the Soviet forces at the front through both partisan warfare and their willingness to endure economic hardship. Massive industrial relocations, using to its fullest extent the Soviet Union's advantage in land area, placed key industries beyond the reach of the Germans. The use of scorched-earth tactics further demonstrated the Soviets' determination to do everything possible to ensure success. Stalin was even willing to forgo certain ideological considerations and appeal to both proletarian and peasant, Communist and nationalist to defend Mother Russia. When the Soviet Union finally emerged victorious from the ashes of World War II, it did so politically stronger than ever before.

Whether Stalin's exploitation of the victory in World War II should be considered shrewd grand strategy or not is open to question. Obviously, he was constantly aware of the peace that he sought. The wartime period is one in which Stalin used every victory of the Red

Army to exact a political price, not from the enemy, but from his temporary allies. Already at the Tehran Conference, by which time the Soviet army had turned the tide on the battlefield, the outlines of Stalin's peace were becoming clear.

Paradoxically, the Soviet Union emerged from World War II simultaneously devastated and more powerful than at any other time in its history. With its new opportunities to negotiate as an equal, the period from the end of the war until 1948 might have brought a shift in Soviet grand strategy from one of isolation from and hostility toward the West to one of greater participation in the international system. Ultimately, it did not.

One of the greatest questions of history has always been whether the wartime alliance against Hitler was doomed to break down in conditions of peace. The assessments usually focus on the series of events that gradually led from collective management of the postwar world to the development of two, hostile camps. Debates on the causes of and blame for the Cold War abound, and there is a sense that if only one side or the other had understood better the fears and concerns of the adversary, the Cold War might have been avoided. Others argue that misperception and miscalculation were as much to blame for the breakdown of the alliance as real events.[16]

But, looking at this issue from the perspective of Soviet grand strategy, the incompatibility of Soviet and Western views of the international system, and of the role of great powers in it, is thrown into sharp relief. Stalin believed that the capitalist powers, given a chance, would seek to destroy the Soviet Union. The "international system" was not the guardian of Soviet interests, it was a threat to them. For Stalin, any cooperation with the West had to be tactical; integration into the international system was suicidal, as this would bring about dependence upon those whose primary goal was the destruction of socialism. There is absolutely no evidence that the alliance experience in World War II had mitigated that view.

Thus, when the Soviets signed the United Nations Charter, which laid out principles of international conduct, and acceded to membership in the Security Council, they did not assume that security interests were to be served by those bodies. Rather, the Soviet Union began building an alternative to the international system dominated by an economically and militarily powerful United States. On the military front, Stalin prepared the Soviet army to fight a great land war, supported by a massive heavy industrial base. The economies of Eastern Europe were forcibly exploited to assist the reconstruction of the Soviet economy, thus extending "fortress Soviet Union" to Berlin.

In spite of the advent of nuclear weapons, Stalin thought that World

War III would be fought just as the previous war had. Nonetheless, he did apparently understand the potential of nuclear weapons, and ordered L. P. Beria, the head of the Commissariat for Internal Affairs, to mobilize the Soviet scientific community into a successful effort to discover the secret of the bomb. It took Stalin's successors to recognize, however, that the nuclear revolution demanded a new strategy. Malenkov and later Khrushchev renounced the "inevitability of war" between the two social systems, and from the mid-1950s there began a new era in Soviet grand strategy; the last major shift until Gorbachev. It freed the Soviet Union to think about a world in which a decisive armed clash between capitalism and socialism might never come.[17]

Even under this new doctrine, known as "peaceful coexistence," the Soviet Union continued to see the international system—dominated by Western capitalism—as fundamentally hostile. Yet there were also new opportunities, for example, to forge alliances with an emerging third world. Soviet strategic thought therefore turned increasingly away from concepts of direct confrontation with the West and emphasized more and more the exploitation of the new nationalism in the developing world. It is often noted that peaceful coexistence was a retreat from the active defeat of capitalism. But this concept had another side; capitalism could also no longer hope to overthrow socialism by military or other "counterrevolutionary" means. Thus, peaceful coexistence, while forgoing war as an instrument of choice, still had at its core a struggle to the finish between two inherently hostile systems.

Khrushchev's third world strategy was attractive for another reason: it allowed the Soviets to extend their power as cheaply as possible, picking up crumbs of support where Western policies had failed. The primary example of this policy was in Egypt. Here, after the United States angered Nasser with the refusal to build the Aswan Dam, the Soviet Union picked up that task and, using Czechoslovakia as a surrogate, began to sell arms to Egypt. A few years later, the first military tutors arrived from the Soviet Union, Czechoslovakia, and Poland.[18] Slowly, the Soviet Union gained a huge foothold in the Middle East, eventually breaking diplomatic relations with Israel and casting its lot with the Arab world. But there is little doubt that the Soviets were, in the first instance, responding to an opportunity, rather than creating their own.

The policy in Africa and Southeast Asia seems to have been little better formulated. African socialism was in its heyday during Khrushchev's reign, in response to the breakup of Western colonial rule. During these years, the Soviet Union enjoyed a considerable reservoir of goodwill as an "alternative" to the West. Again, the Soviets tried to exploit these opportunities cheaply, through military aid and high-visibility economic assistance programs that were often disastrous for the indigenous economies. Little of this accorded with classical Marxist assumptions, but Khrushchev did try his hand at a tortured ideological rationale for supporting these third world revolutions, most of which were headed by personalistic dictators. They were said to be either "socialist leaning" or "on the road to socialism" or "nationalist in character," suggesting that although they might never be real socialist states, they were at least anti-Western.[19]

This policy grew in importance for Khrushchev as the ideological challenge from China intensified during the early 1960s.[20] Determined to show that the Soviet Union was the real leader of the world revolution, Khrushchev supported a variety of questionable clients, many of whom ended up out of power altogether. With the possible exception of Egypt, however, the Soviets seemed to have no long-term plan for *stabilizing* Soviet influence in these countries. More important, when these clients could not keep themselves in power, as was the case in Khrumah's Ghana, the Soviet Union lacked the military power to do anything about it. That fact would be the key difference between these early attempts to exploit "peaceful coexistence" and the third world policy of Leonid Brezhnev.

But the Soviet Union's military weakness in the third world was not Khrushchev's primary problem. The Soviet Union was hardly in a position to secure *itself* in the mid-1950s. By 1957, when Khrushchev had stabilized his internal political position, the Soviet Union's most urgent task was obtaining a viable nuclear deterrent. Stalin had bequeathed the atomic bomb to his successors, and in 1948 the Soviets managed to explode an atomic device some five years ahead of schedule (according to Western estimates) by mobilizing the scientific talent of the country and giving it all resources necessary. The development of the A-bomb, placed under Beria's supervision, was, interestingly, removed from the purview of the professional military. Then, six years after their first atomic bomb, the Soviets exploded a thermonuclear device, but throughout the 1950s they were in the untenable position of complete vulnerability to an American attack. Moreover, American doctrine recognized this condition with a strategy of "massive retaliation" that threatened the Soviet Union with extinction should Soviet military power be used.[21]

Khrushchev, exaggerating Soviet nuclear might on the heels of the success of Sputnik and launching the Soviet Union headlong into several European crises, hardly behaved like the leader of a vulnerable power. Underneath, however, the Soviets were in a furious race to acquire the means to deliver nuclear weapons and hold the United States at risk. The Soviets needed to develop a means of delivery fast and they used an old strategy to do this—mobilizing the resources of the country for a quick, in this case asymmetrical, response to American nuclear might. The Soviets decided that their geographic and technological situation dictated the all-out pursuit of the missile as the means of delivery. They severely curtailed their strategic aviation program after 1959 and, similarly, cut back on the development of both surface ships and submarines. Increasingly, it was the intercontinental ballistic missile (ICBM) program that received the lion's share of resources.[22] This is an instructive episode in understanding Soviet strategic planning. In a centrally directed system, the one real advantage is the ability to mobilize resources and talent to achieve a well-defined goal. The Soviets did precisely this, with the goal of acquiring the most rapidly available means of nuclear delivery, wiping away in about ten years their vulnerability to a preemptive American attack.

The system worked well in giving the Soviet Union the means of delivery that it needed. There were, however, extraordinary strains between the professional military and Khrushchev over the link between the sociopolitical side of doctrine and its military-technical side. From 1954 forward, the Soviet military engaged in a furious debate about the character of war in the nuclear age. There were those who argued that nuclear weapons would not be decisive and thought of ICBMs as ever-larger artillery pieces with longer range and greater yields. Others argued that nuclear war would be unlike any previous armed conflict, obviating the need for ground forces and reducing warfare to one huge exchange at the outset of the war that would be the decisive clash between the two great social systems.[23] Obviously, the implications of the outcome of this debate were enormous. Were one to accept the latter tenets, large cuts in the size of the conventional defense forces could be tolerated and the concerns about command and control and/or war plans would be anachronistic.

Khrushchev interjected himself into the midst of this furious debate, cutting off, prematurely from the point of view of the General Staff, the military's consideration of the nuclear battlefield. Khrushchev fancied himself as something of a military genius, or rather he insisted that nuclear weapons made military genius unnecessary. Unilaterally, he declared missiles to be the decisive element of military power, recommended massive cuts in the size of the armed forces, and abolished the Ground Forces Command. He clearly had his supporters in the military, men like General A. I. Gastilovich, who believed that any war would begin with nuclear strikes and be a very short one indeed. These views, known as "one-variant war," envisioned one massive nuclear blow against enemy territory at the beginning of the war which would decide the outcome.[24]

This interference from the very top also figured heavily in the organizational forms adopted by the Soviet Union to accommodate nuclear weapons. According to the former chief of the Strategic Rocket Forces, V. F. Tolubko, several organizational forms, including a special branch of the air forces (not unlike the U.S. Strategic Air Command solution) and the division of nuclear weapons among the existing services, were considered.[25] But Khrushchev clearly preferred creating a separate service, and it was this option that was chosen.

These decisions then drove other major ones. Khrushchev ordered a cut of two million men from the armed forces, a corresponding cut in the defense budget, and the reduction of all major staffs by one-quarter. Because the chief of the General Staff, Marshal M. V. Zakharov, disagreed with him, Khrushchev fired him. At the same time, Khrushchev was given to quick solutions to exploit the diplomatic potential of nuclear weapons. The most infamous and, ultimately for Khrushchev, the most disastrous of these was the decision to place nuclear weapons on Cuban soil to give the Soviet Union a "quick-fix" deterrent.

On the heels of that debacle, criticism of the general secretary, which had been only thinly veiled in the professional military, started to surface. Two years later, when Khrushchev was ousted for having supported a variety of "hare-brained schemes," that criticism was used by the Soviet military to make an argument for a more systematic means of strategic planning.[26] At the center was the General Staff, determined to return the Soviet Union to a system more like the one that had produced the mechanization of forces in the early 1930s, before Stalin's purge decimated the officer corps.

By 1965, then, the conditions were propitious for the General Staff's bid to "rationalize" military planning in the Soviet Union. Though we do not have access to the high-lever decisions that led to the extraordinary Soviet military buildup between the fall of Khrushchev and, roughly, 1976, it is possible to reconstruct a kind of "strategic plan" or at least "strategic philosophy" that supported it. Post hoc creation is always dangerous, but for purposes of illustration, let us attempt to reconstruct what the Soviet political leaders tried to do in those post-Khrushchev years.

We have some evidence that the catalyst for the buildup was the Cuban missile crisis, an embarrassment of some depth for the Soviet

Union, poised on the threshold of superpower status. There is, of course, the famous statement by Soviet negotiator V. V. Kuznetsov that the West would never be able to put the Soviet Union in that position again. The other evidence is more circumstantial. The decisions taken after the removal of Khrushchev would suggest that his successors understood that the fallacy of "nuclear diplomacy," especially based on meager capability, was its lack of credibility. Khrushchev had gone to all measures of absurdity to convince the West that the Soviet Union was stronger than it was, at one time actually changing the markings on bombers during the May Day parade to give the appearance of a huge bomber fleet when in fact there were only a few aircraft. Satellite imagery would soon render this practice unworkable anyway. But it was less this kind of tomfoolery and more Khrushchev's willingness to take the Soviet Union to the brink of war in this condition of inferiority that worried the Soviet leadership, both political and military.

Brezhnev, the direct beneficiary of Khrushchev's fall, would have non of his dangerous strategy of high-risk foreign policy and meager military power. In many ways, Brezhnev was the most Clausewitzian of Soviet leaders. For him, avoidance of war and preparation for it were inextricably linked. His approach to the international system was to try to reshape it where war was not likely, for instance in the third world, and to engage in cooperation with the powerful states of the system when and where it served Soviet interests.

The twin themes of avoiding war while extending power cautiously were captured in the Soviet version of detente. The Soviets believed that détente was not a Western policy choice but an "objective condition" in which the West recognized that the correlation of forces, a kind of measurement of how history is progressing, was turning in the Soviet Union's favor.[27] It followed that the stronger the Soviet Union was, the more cooperative other states would be. This was an interesting change from Stalin, who had viewed the international system with dread and tried to create a fortress, first in the Soviet Union and then in Eastern Europe, and from Khrushchev, who sought to develop an active foreign policy and to shield the Soviet Union by blustery and largely empty nuclear threats. Brezhnev understood that the protection of the homeland and the projection of power were necessarily linked.

Brezhnev believed that proper preparation of the country for war, a prerequisite for the avoidance of war, should give the Soviet Union a range of military forces, both nuclear and conventional. Those forces could, in turn, be used to give the Soviet Union equal status with the

United States, to exploit weakening capitalism, and to give the Soviet Union a broad role in the world that it once, because of its dire vulnerability, pathologically feared.

The philosophy underlying the defense buildup is perhaps best understood as having two dimensions. On the first dimension, preparing the country for wary, Brezhnev once declared, "Let no one, for his part, try to talk to us in terms of ultimatums and strength. . . . We have everything necessary—a genuine peace policy, military might and the unity of Soviet people—to ensure the inviolability of our borders against any encroachments, and to defend the gains of socialism."[28] In short, under Brezhnev the Soviet Union would achieve Stalin's dream of territorial invulnerability. There was also, however, a diplomatic dimension, since Brezhnev's view of detente with the West contained an active foreign policy. In his view, Soviet military power was now so great that the West had to accept the Soviet Union as an equal. Because of that, it was entitled to a role in all parts of the world.

This then was grand strategy as understood by the Soviet leadership under Brezhnev. The Soviet Union was willing to devote extraordinary resources to fulfilling its tenets. "Military influence" is often an explanation given for huge defense budgets. But in the Soviet Union, where just a few years before the military institution had been powerless to stop Khrushchev's reckless slashing of their resources, it is hard to posit institutional power as the reason for Soviet defense spending. The alternative explanation is that the political leadership saw military power as the most important objective of the state and undertook to build those forces.

The "golden age" of Soviet strategic thought, with its twin aims of securing the homeland and extending power abroad, did not last long. The Soviet Union did achieve recognition as a superpower and found and supported clients from the Horn of Africa to Central America. On the other hand, events such as the Camp David Accords called into question the bold statement that the Soviet Union would always have to be consulted in all important diplomatic efforts. But, on balance, had Brezhnev's reign ended in 1979, when a victorious outcome of the invasion of Afghanistan seemed foreordained, his foreign policy would most likely have been judged a spectacular success.

By the time of his death in 1982, though, Soviet grand strategy had run aground. The Soviet Union faced challenges on all fronts. More than one hundred thousand Soviet soldiers were mired in a hopeless war in Afghanistan. That war and the perception of the unchecked growth of Soviet influence helped to bring Ronald Reagan to power. In

Reagan, the Soviets confronted a leader dedicated to an enormous buildup of American military force and willing to reassert American power.

The Soviets, on the other hand, possessed an economy in crisis, stagnant and structured for the demands of the 1930s, when production of steel was the measure of economic might. Moreover, the Soviets found that the revolutionary movements that they had helped to bring to power had simply become weak, anti-Western states that needed economic assistance which the Soviets could not afford to give. There was, in some sense, a socialist alternative to the international system led by the West, but it was populated by states whose economies were worse off than that of the Soviet Union itself. The drain on Soviet economic resources and the political costs were mounting. Brezhnev had been wrong: an increasingly strong Soviet Union, determined to extend its power at the expense of the West, could not simultaneously count on the cooperation of the United States and its allies. It was possible to be too powerful militarily and for that very condition to weaken one's strategical and political objectives.

Thus, when Mikhail Gorbachev took power in March 1985, he confronted a failing foreign policy just as surely as he faced a failed economy. The two were also linked because it was clear to Soviet reformers that the economy would never recover in isolation from the trade, credits, and technological know-how that the international system could provide. Yet a Soviet foreign policy that so clearly challenged the interests of economic giants like Japan, the United States, and China only deepened that isolation. Gorbachev, in his first major address to the Party Congress in February 1986, called for "new thinking" in foreign policy to parallel perestroika at home.[29] Although the tenets of the policy were slow to take shape and seemed at the time hollow rhetoric, in later speeches he talked of "mutual security" replacing unilateral means; of interdependence, not isolation; and of humankind, not class interests. As the policy has been played out, it can be argued that the "new thinking" is, in fact, an important shift in Soviet thinking about the international system.

The core belief in the protection of the Soviet Union as the key has not changed since the days of Stalin. But, under Gorbachev, a sense that security need not be bought by military power alone has begun to emerge. Gorbachev's situation is fundamentally different than that of his predecessors, and for that he can thank Brezhnev. It is really true that the Soviet Union is so strong, militarily, that exploitation, blackmail, or direct attack by any hostile power seems remote.

On the other hand, the paradoxes of the military buildup, the paradoxes lost on Brezhnev, inform the "new thinking." First, the economic costs of the buildup are clear and are increasingly coming under fire. Much has been written in the West about the economic logic of the new thinking.[30] But it is vital not to lose sight of the political logic of the new thinking, and it is in understanding that logic that the key to the important departure from the past can be found. The primacy of politics has always been a phrase in Soviet lexicon, but according to Soviet critics of Brezhnev, the phrase was never fully understood.

Numerous examples are now cited: the failure to see that the decision to deploy the SS-20 intermediate-range nuclear missile would provoke NATO's counterdeployment; the inability to see that military power was unlikely to bring political victory in Afghanistan; the mistaken notion in the 1970s that there would be no backlash to the buildup of Soviet military power.[31]

The Soviets are searching to understand how politics and military power became divorced. Some point to the now discredited Brezhnev as one who revered military power. Pictures of "Marshal of the Soviet Union" Brezhnev, bedecked in military medals that he did not earn, are now sources of ridicule. Others point to the decision-making system. They argue that foreign policy was made by a small clique with a militarized view of the world. This view was supposedly reinforced by a system that valued narrowly military advice in isolation from political common sense.[32]

But increasingly the debate has revealed interest in more fundamental and fascinating questions. For the first time, the Soviet Union seems ready to link its fate permanently to that of the international system that it has continually denounced as exploitative. One could argue that, in the military sphere, the advent of nuclear weapons makes such linkages a fact of life, not a policy choice. "Mutual security" is a phrase that resonates, but the Soviets are still determined to build the very best military forces that they can, within budgetary limitations. Soviet calls for complete disarmament belie the tough negotiating positions that seek to protect Soviet advantages while diminishing the threat from the West. Yet offers to unilaterally reduce their military forces seem to suggest tacit acceptance of what the West has said all along: Soviet military power is in excess of what is needed to legitimately defend the Soviet Union.

It is possible to argue that mutual security is nothing more than another tactical retreat from the ultimate goal of military superiority. The Soviet economic and technological base is too weak at present to sustain another prolonged military buildup. There is plenty of evidence to support this position. Elements of the Soviet military, long before Gorbachev, were vocal about the need to restructure the technological and economic base in order to meet the demands of the battlefield of the twenty-first century.[33]

On the other hand, this may be a very long tactical retreat in any case. The Soviet economy will not be reformed quickly, and the accumulating evidence is that the goal now is to build a stronger civilian base for the economy. That base can in turn serve military goals, but the priority that the military has had over skilled labor and scarce resources is being curtailed.[34]

That realization is made more interesting by the emphasis in the Soviet debate on the growth of interdependence among states. The Soviets are beginning to ask whether their own fear of, and isolation from, the international system is responsible in large part for the tensions that the Soviet Union faces with the rest of the world. They may be ready to admit that the belief in the international system's desire to destroy socialism took on the character of self-fulfilling prophecy. The "go it alone" strategy, which always saw cooperation with the great powers as tactical, was most certainly costly. But some Soviets now argue that, whatever its costs in the past, it is simply suicidal in the modern era.

If the fact of interdependence is carried to its logical conclusion, the Soviet Union would have no choice but to abandon economic isolation. The Soviets have made known their desire to join the General Agreement on Tariffs and Trade and their future interest in the International Monetary Fund, institutions that they once called tools of capitalist exploitation.[35] Moscow must know that the structure of its economy is far from compatible with these international institutions. But the desire to join them speaks volumes if one compares this to the immediate postwar period, when fear of Western exploitation through economic assistance led the Soviets to reject the Marshall Plan. The exchange of economic information and the role that the IMF would play in the Soviet domestic economy are far greater than the Marshall Plan demanded.

Finally, there is the matter of the primacy of humankind interests over class interest. At the core of the Soviet isolation from the international system has been the belief that the Soviet Union is different from the capitalist states that dominate that system. Although ideology has not governed Soviet international behavior, it has provided a framework, a way of viewing the world. Thus, when the Soviet Union sought friends, it looked to those states that were disaffected with the international system. The Soviet Union proclaimed itself to be in favor of the disenfranchised nations and peoples of the world and, when it could do so without fear of war, acted on their behalf.

This fundamental concept survived Lenin, Stalin, Khrushchev, and Brezhnev. Cooperation with powerful states of the international system was never ruled out. In fact, it was pursued when it served Soviet interests and abandoned when it did not. But if the final victory of the proletariat often became a distant echo, it was at least audible. Slowly, most notably in the speeches of Foreign Minister Eduard Shevardnadze, the defense of class interests and the victory of the proletariat has disappeared. Replacing it are concepts of national interest that reflect no tension at all between the interests of the proletariat and those of the extant international order. Perhaps it was this that led Yegor Ligachev to say that all of the talk about "humankind interest" was "confusing our friends."[36] In fact, the Soviets may finally be on the verge of resolving the confusion in their policies toward the outside world that they inherited when the worldwide victory of socialism did not obtain in 1917. Indeed, its more radical internal reforms (abandoning the Soviet Communist party's monopoly of political power, encouraging moves toward a "market" economy) suggest that the time may soon come when even that most basic assumption of the Marxist-Leninist philosophy—that the "socialist" Soviet Union was locked into an unending battle with the "capitalist" outside world—might also be cast aside. In that case, it would become a "normal" great power (unless it disintegrates internally), with a "normal" sort of grand strategy.

But the resolution of the "dual identity" of Soviet grand strategy—if that is what it is—is not cost-free. The Soviet Union's great influence in the world has been based on opposition to the interests of the powerful and support of the weak. Military power gave the Soviet Union a special function to play in the grand drama of decolonialization and brought a few new socialist states into existence. Gorbachev has yet to articulate an organizing myth for Soviet foreign policy to replace the victory of the proletariat.

If the Soviet Union is to become, instead, a "normal" state, seeking to wield its influence among the powerful, it will need to find new instruments, because those states will not be easily intimidated. In this regard, the emphasis on political means may take on a life of its own and the role of military power may never be quite the same. This is not to say that military power will have no part to play; it is to say that the primacy that it has had in support of the Soviet Union's goals will have to change.

The greatest lesson of the Brezhnev period was not that military forces cost money; it was that threatening other powerful states with military extinction is not credible, and is also counterproductive. The Soviet Union was respected and feared in the international system; it was also ruled out as a suitable long-term partner for cooperation.

But it will undoubtedly be easier to abandon the old policy than to make the new endure. Facile diplomacy, an enormous strength of Gorbachev era, will not suffice alone. The Soviet Union needs the other traditional elements of power, most importantly economic strength, in order to play this new game. The Soviet Union has very few cards to play. The new foreign policy was, in large part, born to support perestroika at home. Ultimately, though, the relationship may be reversed; the new foreign policy cannot survive long unless perestroika gives the Soviet Union the economic clout to make it a success.

The United States and Grand Strategy

10

American Grand Strategy, Today and Tomorrow: Learning from the European Experience

Paul Kennedy

All great powers are unique, for the simple reason that they differ from each other in both time and place. The Victorian Britain which claimed to be bringing a "Pax Britannica" to the world was as distinct from the Roman Empire to which it made comparison as it was from the United States that, a century later, was often regarded as having established its own "Pax Americana." In geography, constitution, population, culture, and place in world history, the differences between all three powers were immense. And yet, in the implementation of grand strategy, all of them—as well as many other great powers, from the Ottoman Empire to Soviet Russia—faced the same tests and problems.

What were—and are—those tests and problems? Essentially, they have to do with the search for "security," broadly defined, in both wartime and peacetime. In an anarchic world that lacks a single sovereign power to order its destinies, tribes, cities, empires, and nation-states have jostled alongside each other, and all too often gone to war with each other for a whole variety of motives—for land, for trade, for gold, because of dynastic or religious or ideological rivalries, out of fear of being overtaken or a desire to overtake. Not surprisingly, therefore, the search for "security" has usually been seen in *military* terms—of being strong enough to deter potential foes or, if that fails, to defeat those enemies on the field of battle, thus preserving the tribe's (or the empire's) existence and interests.

But beyond this narrowly military conception of security—most easily seen in the relatively recent creation of "security" advisers to premiers and presidents, whose task is to advise upon the roles and policies of the armed forces—there is an altogether broader conception. It is, in a curious way, very American: it is about the implementation of policies which would secure (in the founding Fathers' words) "life, liberty, and the pursuit of happiness" for the polity in question, however restricted that polity might be.* It is a conception closely related to the larger strategic ideas of Clausewitz and Liddell Hart, which were described earlier in this volume: "To begin with, a true grand strategy was now to do with peace as much as (perhaps even more than) war. It was about the evolution and integration of policies that should operate for decades, or even for centuries. It did not cease at a war's end, nor commence at its beginning."[1]

It is because of the essentially *political* nature of grand strategy— What are this nation's larger aims in the world, and how best can they be secured?—that there has to be such a heavy focus upon the issue of reconciling ends and means. There was, for example, little sense from Olivares's perspective in devoting all of Spain's resources in the 1630s to the defeat of the Netherlands, if that meant withdrawal from other important parts of the Habsburg Empire.[2] There was a deep concern, and necessarily so, by British politicians in the early decades of the twentieth century, that if they committed a disproportionate share of the nation's manpower and resources into a European "continental commitment," they would denude their imperial defenses and weaken their entire economy; yet if they did not strive by all means to preserve the European balance of power, German hegemony was virtually inevitable.[3] Since the beginning of the atomic age, there has been a repeated questioning of the utility of "prevailing" in an all-out nuclear war if, say, the Northern Hemisphere was devastated by the effects of that fighting. Indeed, the coming of nuclear weapons provides the strongest possible reinforcement to Liddell Hart's earlier observation that since "the object in war is to obtain a better peace—if only from your own point of view— . . . it is essential to conduct war with constant regard to the peace you desire."[4] By the same token,

---

*The term *polity* is important here. No one would suggest that the Ming dynasty, or the Tudor monarchy, or the Soviet Politburo was primarily intent upon aiding the liberty and happiness of its subjects (that is, most of the population), because the latter were not *included* in the "polity" of the country in question. In a democracy, the aims remain the same, but the constituent "polity," and therefore the means required to achieve those aims, are enormously enlarged.

presumably, it is essential to conduct peace with constant regard to the war (or wars) that you may be called upon to fight.

This is not to say that grand strategy in peace is identical to grand strategy in war. Clearly, the latter condition calls for "blood, sweat, toil and tears" to a degree that simply does not exist in peacetime. Like an amateur athlete, the nation-state in times of peace (that is, while not engaged in outright physical competition) has to balance many desiderata—earning its keep, enjoying its pleasures, and keeping fit and strong; but when the race (or a conflict) occurs, a far larger amount of energies and effort is given to winning and fighting, and the other elements are left until later. After the event, one can always return to "normal."

Grand strategy in war is, therefore, necessarily more military than it is in peace.[5] The real task for the polity in question is to ensure that, in wartime, the nonmilitary aspects are not totally neglected (a failing of Germany in both world wars); and that, in peacetime, the military aspects are not totally neglected (a temptation to which the publics of the post-1919 democracies, recoiling in shock from World War I, were particularly prone). Once again, it is a question of not becoming totally unbalanced.

But if the wartime task of balancing ends and means also exists in the peacetime execution of a nation's grand strategy, there is the additional problem that politically it may be the harder to achieve, year after year, since the conditions of peace conduce to turning the polity's attention to other priorities and activities. This is especially so in a period of prolonged peace, or when a nation's military security is being chiefly provided by another. It is easy to declare that a balance must always be struck between devoting too little to defense upon the one hand, and too much to defense on the other; yet the variables which may affect the achieving of a balanced national "security" are so many, given the fluctuating course of international affairs, that very few polities have succeeded over time in preserving the right equilibrium.

It is the risks of allocating too *little* to defense which have occupied the more prominent place in the minds of Western politicians and strategists during the post-1945 decades, and for very obvious reasons. They had strong memories of how, in the aftermath of World War I and under the then-prevailing political assumption that it had been "a war to end all wars," the Western democracies had steadily run down their armed forces, frozen a great deal of weapons development, concentrated their attention upon domestic affairs, allowed the decay of former alliances—and thus placed their countries in the weakest possible position when the revisionist, Fascist states began their assault upon the international status quo during the 1930s. Militarily vulnerable, diplomatically divided, distracted by internal political and economic issues, the democracies were simply unable to institute a grand strategy which (to employ Earle's definition once again) "so integrates the policies and armaments of the nation that the resort to war is . . . rendered unnecessary."[6] Given that these sorry "lessons of appeasement" were being imbibed after 1945 by a successor generation of Western statesmen deeply worried about the threat posed by Soviet military power, it was perfectly understandable that they conceived their chief political task as being to warn their electorates against insufficient spending on defense, and about the need to stand firm against aggression.[7]

It is worth recalling, however, that among an even earlier generation of Western statesmen the greater concern was that too *much* of the nation's resources, personnel, energies, and attention would be devoted to military purposes and to the pursuit of victory in the field at all costs—to the detriment not only of the country's economic future, but also, and perhaps even more importantly, of its liberal political culture. Perhaps the best examples of this concern can be witnessed among British politicians and thinkers—predominantly in the Liberal party, but also including many traditional Conservatives—as that country debated how to respond to increased international pressures after 1890, and in particular to the rising challenge of imperial Germany. The "responses" called for by the British right—the abandonment of free trade and a return to protectionism, large-scale increases in the defense forces, the introduction of conscription, controls over aliens and other "subversive" figures, and unflinching commitment to all of Britain's overseas obligations—greatly alarmed the traditionalists. Of what real use was enhanced military power, they asked, if it came at the cost of the nation's civil liberties, of its laissez-faire political culture, of its economic well-being? Could Britain compete with the protectionist, autocratic, militarized garrison-state of Prussia Germany only by becoming a "garrison-state" itself?[8] To some degree, that debate foreshadowed many of the questions asked by Eisenhower in the 1950s, and by later critics not only of the so-called military-industrial complex but also of the tendency toward limiting individual and press freedoms, and distorting American industry, technology, and science in pursuit of "winning" the Cold War against the Soviet Union. Did the spiraling demands for *military* security mean that the American polity would be compelled to ignore or downgrade the needs of *social* security, *educational* security, *fiscal* security, and even *environmental* security—and at what later, longer-term costs?[9] On the other hand, if there were insufficient armed forces to deter the Soviet Union and to deal with regional security threats (such as Libya and North Korea), would not the nation's interests— and the physical security of its citizens—be seriously hurt?

Both in the Britain of Edward VII and in Eisenhower's America—or, for that matter, in Gorbachev's Soviet Union and Briand's France—the pattern was the same: the debate was conducted (and settled) at the political level; it involved an acute awareness of the "trade-off" between ends and means and between the nation's more immediate needs and its longer-term requirements; and it involved the recognition that grand strategy in both peacetime and wartime involved much more than purely military policy.

But if these are the "tests and problems" facing all great powers as they strive to effect a balanced grand strategy in peacetime, it nonetheless remains true—as I mentioned in the first paragraph of this essay—that each power is different in time and in place. Before any attempt is made to measure American grand strategy against the general principles and European experiences detailed above, therefore, it is necessary to understand the unique historical circumstances in which the decision makers of the United States have had to operate.

Since its early decades, the United States has been the beneficiary of a cluster of highly favorable geographical and technical factors. Protected by the Atlantic Ocean—and the Royal Navy—from serious external threat, the nation could divert its energies from swords into ploughshares throughout most of the nineteenth century. Rich in raw materials and food, but relatively sparse in population, it possessed resources which could only be properly exploited by the introduction of labor-intensive machinery, which thereby gave its entrepreneurs an advantage over foreign rivals; and by a communications revolution (steamship, railway, telegraph) which vastly enhanced its position in the global marketplace. Moreover, while it held aloof from Europe's political and military quarrels, the United States always benefited from an enormous two-way economic relationship: sending out vast supplies of cotton, timber, beef, and (later) machinery, and importing early European technology and large sums of capital to boost its own industrialization. It did not grow up in a vacuum.

What the wars of 1914–1918 and 1939–1945 did was to accelerate two broad trends in global politics. The first was to harness—and boost—the still-enormous potential of the North American continent for both peaceful and military purposes. The second was to weaken the economies of the European great powers by diverting too much of their limited resources into armaments, and then into mutually devastating wars.[10] As one demicontinent fell, another rose; and the wars quickened the pace of change.

After 1945, therefore, the United States found itself in a remarkable position (in a way, somewhat equivalent to Great Britain's in 1815). It possessed more than half the free world's manufacturing capacity, it had heavily invested in new industry and products, and it managed to channel its prodigious wartime energies into an equally impressive peacetime growth during the years following. Since all of its former commercial rivals had been ruined by the war, the United States enjoyed a near monopoly in domestic and in many foreign markets. Not surprisingly, this preeminent position confirmed in the minds of many, perhaps most of its citizens the superiority of "the American way": free markets, a low level of government intervention, a constitutional separation of powers, a Protestant "work ethic," and so on.[11]

It was in these favorable circumstances, with the United States having become clear number one in world affairs and its polity blithely assuming that all other societies from India to Hungary would wish to imitate its own laissez-faire and democratic practices, that Washington found itself in a "cold" war with the Soviet Union. Rejecting any return to post-1920 isolationism, the American leadership took up the Soviet challenge at every level, military, diplomatic, economic, and ideological. After 1947, the USSR was to be "contained" by a ring of alliances, treaties, and bases stretching from Norway to Japan. Economically, the battered societies of Europe and Japan were to be restored to full health within the Western capitalist system by the injection of vast amounts of Marshall aid and by the erection of a strong international framework for "open" trade and investment. Ideologically, a struggle would be waged to win the minds and hearts of other peoples, and to lead them away from the follies of communism. To be sure, this more active role involved large costs—in foreign economic aid, in defense spending, and in human lives (especially with the advent of the Korean War); but those were costs which Americans were now both willing and able to pay.

Four decades later, and admittedly with the benefit of hindsight, the observer cannot help being impressed by how successful on the whole that basic American grand strategy has been. The framework of an open world trading system permitted an extraordinary expansion in global manufacturing, commerce, and investment[12] and raised the standards of living of millions of people to unimagined levels. The Soviet military threat was "contained" throughout the greater part of the world, and the chief alliances (NATO, U.S.-Japan Defense Treaty) have held firm. In the realms of political culture and ideology, Marxism has made very few long-term breakthroughs. In sum, if Truman, Marshall, Acheson, and their advisers had been asked to describe what sort of world order they hoped would be in place forty years later, the broad outlines might look very close to what exists today.

Nonetheless, the world has changed a great deal since the early years of the Cold War, and in ways which could significantly affect the evolution of long-term American grand strategy. Such changes suggest that, at the least, the component parts of the post-1945 strategy be reexamined to discover whether they need readjusting in the light of these global transformations—and whether the order of policy priorities also requires amendment, to permit the national strategy to be recast upon stronger foundations. Such a reexamination, far from being radical, is inherently *conservative* in nature, on the lines of Lord Salisbury's famous remark that the commonest political error was that of "sticking to the carcasses of dead policies."[13]

There are three, perhaps four aspects to American grand strategy which, it may be argued, have significantly changed since 1945. The first is the ending of the United States' own strategic invulnerability, which for two centuries had given it an advantage that most other powers (France, the Netherlands, Russia) envied but could never hope to emulate. Whereas the advent of very-long-range bombers and intercontinental ballistic missiles makes it difficult, and perhaps impossible, for any nation to defend itself against such assaults, in some ways it was the United States which had lost the most by this development. Moreover, such a transformation was worsened by the almost simultaneous creation of nuclear weapons, at least once the early American monopoly in those technologies had been broken—for the results of an all-out nuclear exchange would render irrelevant the usual geopolitical advantages that the United States had possessed in the two world wars. With thousands of Soviet warheads apparently targeted upon American objects, the nation had dramatically moved into the front line in the event future hostilities escalated into an atomic war.

The second change has been the slow erosion of the United States' undisputed preeminence in technology, manufacturing, and finance—upon which its "rise" as a great power had ultimately rested. Indeed so obvious was this American economic superiority in the first two-thirds of the twentieth century that it had been taken for granted in wartime.[14] In the post-1945 years, American policymakers were occasionally to worry about "over-extending" their commitments and resources;[15] but most of them also tacitly assumed that, given a sufficient amount of time—and in both world wars the United States had been neutral for the first three years of the conflict, and its troops only began fighting in the European theater during the

fourth—the "gigantic boiler" could once again be mobilized to produce a flood of munitions that would succor hard-pressed allies and
overwhelm outgunned enemies. To borrow a phrase from another
time, the massive economic underpinnings to American grand strategy were an "unspoken assumption"[16] of decision makers both inside
and outside the country.

For two (related) reasons, our contemporary global situation no
longer permits that cozy assumption to be made. The first is that, just
as at the nuclear level, the United States now occupies the front-line
position in the conventional defense of its own and general Western
interests, from the Fulda Gap to the Persian Gulf to the Korean Demil-
itarized Zone. In 1917 and again in 1941, the United States enjoyed a
"buffer zone" of thirty-five hundred miles (in the Pacific, of over six
thousand miles) between itself and enemy forces. When it succeeded
to the clear leadership of the Western democracies, after 1945, that geopolitical situation was reversed. As the recent Pentagon-sponsored
Commission on Integrated Long-Term Strategy put it, "Defense planning in the United States has centered for many years on a grand
strategy of extraordinary global sweep. The strategy can be stated
quite simply: forward deployment of American forces, assigned to
oppose invading armies and backed by strong reserves and a capability to use nuclear weapons if necessary. Resting on alliances with
other democratic countries, the strategy aims to draw a line that no
aggressor will dare to cross."[17]

One obvious consequence of this postwar transformation of the
American strategic posture is that it has needed to allocate a far larger
share of its GNP to defense, and to maintain far larger *ready* armed
forces, than was the case in the mid-1900s or mid-1930s. Not only
that, it has spent and maintained much more than its friends and
probable wartime allies. In becoming a "front-line" rather than
"reserve-zone" great power, the United States has—like, say, the
France of 1910—to bear a higher cost for readiness.

Given the revolution in weapons-delivery technology and the
equally plain fact that no other country or countries could have replaced the United States in its post-1945 role as the guarantor of
Western interests, this transformation was presumably unavoidable.
But it is important to take note of it because of the change mentioned
above: that the decisive American technological, manufacturing, and
financial superiority which underpinned its grand strategy for decade
after decade is now no longer so evident. There is no space in this
essay to rehearse the many complex arguments—for and against—
"the relative decline of America."[18] But what is incontrovertible is

that, as compared with, say, Eisenhower's time, its economic position has altered: its technological lead has been eroded in area after area; its manufacturing competitiveness has been reduced, as compared with that in other industrialized countries, and in some sectors (including even military-related) domestic production has been eliminated, creating a dependence upon foreign suppliers; it is unable to pay its way in the world except by borrowing abroad; and in many related ways (educational skills, savings ratios, levels of research and development), its international rankings are not high. The country still possesses considerable strengths, but the point being made here is that it is nowadays much less evidently the "arsenal of democracy" than it was in Franklin Roosevelt's days. One presumes that, under the threat of a future great conflict, many of these worrying trends could be reversed over time (for example, domestic production of various strategic items could be resumed); but—quite aside from the question of whether the conflict would escalate beyond the use of conventional weapons—there remains the fact that, being in a front-line position, the United States is unlikely to enjoy such a breathing space.

The third change, in many respects related to (and the converse of) the second, has been the alterations in the global economic and even strategic balances which have taken place since the early Cold War. In the place of an essentially "bipolar" international system in which the United States and the Soviet Union were militarily far ahead of every other power and together possessed perhaps over 60 percent of total world product, there is now a much different order. Economically, the European Community is as large as the United States (according to some measures, larger), Japan has overtaken the Soviet Union, and East Asia, including China, is undergoing remarkable growth. By many of the traditional assessments of relative economic power— shares of world manufacturing and trade, banking assets, investment flows—there already exists a "multipolar" system, and it appears to be becoming more diffuse from decade to decade. At the same time, even the clear military lead which the two superpowers possessed thirty years ago has to some extent been eroded, with the growth of substantial European-NATO and Communist Chinese armed forces, and the increasing proliferation of nuclear weapons and delivery systems.[19]

To the degree that this "multipolarization" of the international system represents and fulfills the post-1945 American policy of assisting in the recovery of Western Europe and Japan, the trend can be viewed as a positive one. At all events, it is the Soviet Union much more than the United States which has been affected by the change in the overall "correlation of forces" consequent upon the revival of Europe and the rise of China and Japan; and its leadership can gain no comfort from long-term projections that suggest the USSR may possess only the world's fifth-largest economy by early next century.[20] Nevertheless, the United States is also presented with a challenge of its own, that of managing relationships with other important nations of the globe that are of equal or near-equal economic weight, and no longer as dependent upon Washington's favors as was the case three or four decades ago.[21] If, moreover, these alterations in the global balances continue into the future—that is, if the American economy should grow less swiftly than those of Japan and an increasingly united Europe, and if its international indebtedness should intensify—this could imply further and repeated adjustments in intra-alliance relationships, especially if American congressional and public opinion presses for greater "burden sharing" on defense costs. But burden sharing also implies "influence sharing," and that in turn implies a relatively diminished position in international affairs for the United States compared with that which it naturally expected to wield during, say, Kennedy's presidency.

Vulnerability to mass destruction, the erosion of its indisputed economic preeminence, and the growing multipolarity of the international system: those are the three aspects of American *peacetime* grand strategy that have changed the most since the late 1940s.[22] But there may be another, although it is far less certain at this writing: namely, the reduction and possibly the elimination of the threat of communist expansion that has provided not only the "cement" to keep the Western alliances bound together (since World War II), but also the justification for the historically large American defense expenditures and the raison d'être for much of its armed services' operational planning, training, deployments, and force structure. Whether there really could come about an "end of the Cold War," as is nowadays frequently asserted (and hoped for), must remain uncertain for some time to come; a great deal depends not upon anything which the United States can control, but upon the unfolding of events within the Soviet Union and its European satellites. Nonetheless, the concentration by the Soviet leadership upon the critical issue of internal reform—accompanied so far by corresponding efforts to improve external relations and to reduce defense expenditures and force size—makes the global scene far less threatening to Western interests than was the case in 1950 or 1960 or even 1980. Furthermore, the changes which are occurring within the Soviet Union coincide with (and interact with) a profound erosion in enthusiasm for Marxist dogmas across much of the globe; so that, whatever challenges are to be faced by the United States in the years to come, it is difficult to visualize a world divided preponderantly into Communist and non-Communist blocs, as it was often seen to be in past decades.

A reduction in the Soviet "threat" and a decline in the plausibility of a global Communist challenge would not only affect the size and futures of the U.S. armed services;[23] it would also increase the contemporary tendency to raise questions about traditional definitions of "national security." Already there is a widespread conviction in American (and general Western) public opinion that the challenges to military security are much less pressing than those to the nation's economic security—with the Japanese technological and manufacturing challenge being regarded as a more serious "threat" to the well-being of the United States than the Soviet military danger. In addition, attention has swung to the increasing evidence of threats to the environment, of the long-term implications of the weaknesses in the American educational system ("a nation at risk"), and of the damage done to the fabric of American society by drug traffic. Other commentators argue that the federal government's budgetary deficit, and its accumulated international indebtedness, are the most pressing dangers to the country's security. The Iraqi annexation of Kuwait, and the United States' impressive initial response to that action, may have been a salutary reminder that military force still *counts* in an anarchic world, but it is unlikely to suppress the debate that is occurring across the United States about what exactly constitutes national security as the nation enters the increasingly complex 1990s.

All this has implications for the political priorities—and thus the *spending* priorities—of the American polity. The tendency is to argue that, in an age in which the Soviet Union is restructuring its own foreign and defense policies, and in which the global scene itself is becoming much more competitive technologically, resources will need to be reallocated from the U.S. military into other areas (reducing the deficit, dealing with environmental problems, improving infrastructure and education, combating drugs). This is a tendency which the Pentagon will oppose, the more particularly since the political circumstances of the early 1980s allowed it to go ahead and order the prototypes of many new and expensive weapons systems—the full funding for which was presumed to come in later years, an assumption that looks ever less plausible as time goes on.[24]

If the external circumstances have changed greatly since the era of the early Cold War, and if they are continuing to change at a faster pace than before, that poses a more subtle test to American policymakers:

the test of rearranging the different components of their peacetime grand strategy. To at least some extent, the security environment facing the United States as it goes into the 1990s may be likened to that confronting Britain and France in the 1920s. Before World War I, it had been relatively easy for them to implement grand strategy—simply match Germany's expansion, of both land power and sea power, step by step—just as it was relatively easy for Ronald Reagan and Caspar Weinberger to argue that the aim of American grand strategy after 1980 was to match Soviet defense spending. In the altered global political landscape of the 1920s, it was much more difficult for British and French strategists to fix their bearings. The old enemy was less threatening, and might not even be a foe at all. Newer powers were playing a larger role in world affairs. Previous allies were less reliable, more problematic. Moreover, this essentially *military* calculus was itself affected by nonmilitary considerations. Public opinion in the democracies overwhelmingly believed that a great-power war was a thing of the past; weapons were too horrible, the costs too great, for any major conflict to be permitted in the future. Instead, attention was concentrated upon the nation's economic and "social" security: trade competitiveness, financial instability, threats of protectionism, structural unemployment, the needs of education, housing, and health care. Predictably enough, the armed services found it difficult to justify a high level of defense spending in this changed environment; manufacturers moved out of the "military-industrial complex," and many of them withered away completely. This contraction of military spending and industrial "surge capacity" was not, however, attended by a reduction in foreign commitments and obligations. Consequently, when the external security environment became much more threatening again, in the 1930s, both British and French policymakers found themselves in a constrained, uncomfortable position.[25]

This is not, of course, to predict that the United States will find itself in such a vulnerable position around, say, 1997 as the British and French felt themselves to be around 1937. But since it is the aspiration of American planners to design a strategy "for the long term" and "for a wide range of conflicts," it is worth asking whether the contemporary wisdom[26] has got the balance right.

At the moment, there are still too many signs that most American "strategists" still think in too narrowly *military* terms: arms control, out-of-area operations, "smart" weapons, large versus small "platforms," ballistic missile defense, procurement, manpower, and so on, occupy the center of their mental stage. The danger is that, as perestroika and détente impinge upon the agenda of policy priorities, and upon congressional and public opinion, such a heavily military level of analysis might seem to be anachronistic and irrelevant— leaving Pentagon planners almost as bereft of their bearings as British and French General Staffs in the 1920s.

If the more immediate problem is to avoid a too narrow and conservative approach to "grand strategy" on the part of the United States government, the longer-term challenge is to prepare the country to be in a reasonably strong and flexible position to operate in the unpredictable and perhaps volatile circumstances of the late 1990s and early twenty-first century—which brings us back to the diplomatic and the economic and all of the other elements of any properly integrated long-term strategy in peacetime.

1. By far the most important aim of American grand strategy today, and into the foreseeable future, has to be the avoidance of nuclear war. In the profoundest way possible, the coming of atomic weapons has transformed the strategical landscape, since they give to any state possessing them the capability of mass, indiscriminate destruction, even of mankind itself. It was for this reason that Bernard Brodie wrote, as early as 1946, that "thus far the chief purpose of a military establishment has been to win wars. From now on its chief purpose must be to avert them."[27] It is an observation of even greater truth nowadays, when both of the superpowers possess literally thousands of warheads, with a capacity to blow up the world. If, as Professor Robert Jervis nicely puts it, "a rational strategy for the employment of nuclear weapons is a contradiction in terms,"[28] the overriding task of statecraft is *nuclear prevention*, with all of its consequential implications.

Among those implications, one would expect to see a tremendous attention being paid to the technical aspects of "command and control," to prevent the possibility of an *accidental* nuclear war, or an unintended (or unauthorized) nuclear strike escalating into a far more destructive exchange.[29] The second implication is the need for the total abandonment of the older NATO doctrine of planning to use nuclear weapons in the event of a Red Army "breakthrough" in Central Europe during a conventional war, a possibility made increasingly implausible by the withdrawal of Soviet ground and air forces from much of East-Central Europe; and for the restructuring of the types of nuclear weapons held by the West, so that the emphasis is on those "that are good for retaliation but not for initial attack, that can survive a first strike by the Soviet Union but are poor instruments for a disarming first strike against it."[30] The third implication is a massive "build-down" in the overall number of missiles and delivery systems, to an irreducible minimum required for deterrence—on the assumption that neither of the superpowers would abandon *all* their weapons, given the fear of nuclear blackmail and the proliferation and development of the nuclear capabilities of an increasing number of third countries.

The fourth and final implication flows from the third. It is the need for the United States and the Soviet Union to work together to arrest nuclear proliferation—something that is in the interests of both powers—as well as to combine in the effort to persuade the other nations with nuclear capability (France, Britain, the People's Republic of China) to agree upon multilateral arms control and verification. There are, indeed, good arguments for the United States and the Soviet Union to work together on many other fronts, from environmental issues to space exploration to settling some of the world's regional conflicts; but the clearest and most worthy would be in measures to halt nuclear proliferation.

2. The second aim of American grand strategy should be to create armed forces flexible enough to deal with a variety of possible fighting contingencies. It is an inordinately difficult task for a global superpower like the United States, because of the types of war it might be called upon to fight. When a large-scale conventional conflict against the Warsaw Pact was the most probable scenario, then the funds needed to be spent upon tactical aircraft, main battle tanks, and the like. Since it has become likely that the United States and the Soviet Union will avoid a direct clash but that there will be more activity in the third world, the weapons mix has to be different: small arms, helicopters, light carriers, plus an enhanced role for the U.S. Marine Corps. Here again, the American dilemma is not unlike that faced earlier in this century by British planners, whose ground forces were supposed to be capable of fighting on the northwest frontier of India *and* in continental Europe—and that at a time (after 1920) when funds for the armed services were being curtailed.[31] If there is a "lesson" from that experience—which led to the virtual elimination of the British army's capacity to fight in Europe, and then desperate, belated attempts to recover that capacity at the end of the 1930s—it is that forces and weapons which are flexible enough to operate in various battle scenarios are the ones to be preserved in periods of austerity. (Helicopters are more useful than main battle tanks; Stinger missiles are more multipurpose than AEGIS-type cruisers). This is much easier said than done; but unless the general principle is ad-

hered to, military cutbacks might lead to the elimination of those more flexible weapons systems in favor of expensive, single-purpose projects—and they almost certainly would lead to the elimination of items useful for interservice cooperation. Throughout this century, air marshals have preferred strategic bombers to ground-support planes, and admirals favored battleships over landing craft.

3. The third aim and feature of American grand strategy stands in close relationship to the second: it is to preserve, and possibly to some degree to redefine, the system of alliances which the United States has constructed across the globe since the late 1940s. They are no less important now than they were at the height of the Cold War, and as the final decade of this century approaches, they offer both risks—of erosion and weakening in the face of Gorbachev's détente policies and of inter-allied disagreements upon defense "burden sharing"—and opportunities—of a certain reallocation of roles and forces as the American contribution to the protection of Western interests is reduced, while that of some of its more prosperous partners (Japan, NATO Europe, Australia) is increased.

This aim also is easier stated than achieved, for two reasons. The first is that the very coming of a period of improved relations with the Soviet Union takes away the "cement" which has kept the American-led system of alliances together, while at the same time helping to increase the impression that the needs of "military security" are of much less import than those of "economic security." It thus places the burden of proof upon those American officials and diplomats who are attempting to persuade reluctant allies to spend more on defense at a time when the United States is planning to spend less in real terms and when the East-West scene is relatively tranquil. Here, too, there are distant echoes of the 1920s, when British politicians strove, with little result, to persuade the self-governing Dominions to increase their contributions of money and manpower to the common defense.[32] Yet, however difficult the prospect, the issue of burden sharing between the United States and its allies needs to be pursued, simply because it will not go away. It is not a good strategy for the long term to have a country with a five-trillion-dollar economy contributing so much more for defense in order to protect allies possessing economies of six trillion dollars (the European Community) and three trillion dollars (Japan) from the threat posed by a country with an economy of little over two trillion dollars (the USSR).*

*All figures are rounded off and approximate.

The other reason why this goal is difficult to achieve in practice is that an alteration in burden sharing will most likely bring with it some changes in influence sharing. Yet, psychologically, the leading nation in an alliance usually finds it difficult to agree to a reduction in its power to influence events.[33] Even as the United States turns, say, to Japan to offer loans to third world countries which Washington itself, facing its budget deficits, cannot now provide; or as it urges the Federal Republic of Germany to take the lead in assisting the economic development of Eastern Europe—even so, it will find it hard to accept the diminution of influence that such measures imply.

Given this emphasis upon redefining relationships with American allies at a time when vast changes are occurring in most Communist societies and when many parts of the globe (but especially East Asia) are undergoing remarkable transformations, there is a clear need for the United States polity to understand much more about what is going on outside its borders. On the diplomatic front, this implies not only a considerable enhancement in the position of the State Department and the Foreign Service, but also a massive increase in educational and media coverage of international events. In what is forecast to be increasingly a "knowledge-driven" society, ignorance of foreign societies, cultures, and languages is likely to prove a serious strategic weakness.

4. The final aim being suggested here is probably the most important of all, excepting the need to avoid major war itself. It is to institute serious measures to reverse those trends which have already begun to weaken the preeminent position in the world that the United States occupied three or four decades earlier. From a grand-strategic viewpoint (not to say moral, social, and human viewpoints), it is worth asking whether it is not unhealthy and alarming that the country's national debt should be growing so rapidly in peacetime; that the United States should have transformed itself from being the world's greatest creditor nation to being (by some measures of accounting) its greatest debtor, with a heavy reliance each month upon foreign purchases of Treasury issues; that it has permitted the erosion or even collapse of American industry in certain key strategic sectors, so that it is now dependent upon East Asian suppliers for the electronic innards of so many of its own weapons systems; and that the educational and skill levels of its work force are below those in virtually all other advanced societies, which in turn reflects the crisis in the American inner-city schools, educational standards, and national efficiency.

Although the answer to those questions can only be affirmative, the challenge is to implement reform measures precisely in a period when the U.S. government's fiscal position is precarious, and therefore unencouraging of investment in domestic needs; and when its external position, although challenged by critical regional problems (especially in the Middle East), is significantly better in *global* terms than it appeared to be at the beginning of the 1980s. In such circumstances, it may appear an anachronism to worry about whether the United States nowadays has the manufacturing "surge capacity" to supply the increased needs of its armed forces in the event of future protracted and major wars. It may also seem unreal to express concern about the American defense industry's reliance upon foreign-made products (such as microprocessors) when it is so easy to ship or fly them across the Pacific. It may look redundant to agitate about the United States' deep reliance upon the continued inflow of foreign funds at a time when capital flows seem to respond overnight to technical encouragements such as an increase in interest rates.

Perhaps those concerns *are* anachronistic and redundant. Perhaps the international system of states that exists in this post-Hiroshima era is one in which great-power wars will never happen again, and international cooperation and integration (albeit with lots of grumbling, and continued regional conflicts from time to time) will be the order of the day. Perhaps the nation-state itself is an anachronism, and "actors" such as multinational corporations and international financial institutions have eroded much of the state's former autonomy. Perhaps. But those responsible for creating a country's "integrated, long-term strategy" cannot afford to make those assumptions, or to use them as the basis for national policies. It may be that foreign capital will continue to flow into the United States—though it is worth recalling that that was the general assumption about *American* capital pouring into Europe in the mid-1920s, until the global economic crisis occurred. It may be that foreign-made products and weapons parts will continue to flow across the Pacific, to meet the needs of the American defense industry—though it is worth recalling that a similar situation existed in Edwardian Britain, which consequently found itself acutely embarrassed in 1914 to discover how reliant its defense manufacturers had become upon German-made products (ball bearings, aircraft engines, dyestuffs, optical equipment, and so forth). It may be that there is no need to worry about an American industrial surge capacity because there will be no more major wars—though that, too, it is worth recalling, was also the prevailing belief of the Western democracies in the 1920s. All these issues reduce—or increase—a country's strategical vulnerability.

In any case, whether there will or will not be great tests of the

United States' military effectiveness at some future date, it simply is not a sensible strategy to have the leading power of the Western world resting upon financial and industrial foundations that are increasingly less competitive internationally than they were in Eisenhower's period, while at the same time the same nation has retained (indeed, expanded) all of the military and political commitments of those early post–World War II years. If grand strategy is about reconciling ends and means, it is worth taking notice if the ends remain the same but some of the means are relatively diminishing.[34]

All of the above points to the inordinate complexity, in peacetime perhaps even more than in wartime, of managing all of the variables that must be brought together in order to carry out an effective, long-term grand strategy; no wonder that Clausewitz described it as an art, not a science. But since it is not humanly possible to prepare for everything that may happen in the unpredictable and turbulent world of the early twenty-first century, the task is to structure the armed forces, and the economy and society upon which they rest, to be in a good position to meet contingencies. In other words, the United States ought, while seeking to fulfill its people's peacetime desires, to maintain a reservoir of productive and financial and technological and educational strength—so that if a "1920s" world unfortunately turned into a "1930s" world at some point in the future, the nation would not *then* discover that its grand strategy was crippled by a whole series of defense "deficiencies"[35] which a faltering economy could not easily correct.

Notes

Chapter One. Grand Strategy in War and Peace

1 Carl von Clausewitz, *On War*, edited and translated by Michael Howard and Peter Paret (Princeton, N.J., 1976), pp. 127–132; John R. Elting, *The Super-Strategists* (New York, 1985), p. 2, quoting from a U.S. Military Academy handbook.

2 Archer Jones, *The Art of War in the Western World* (Urbana, Ill., 1987), p. 55.

3 "Operational military activity involves the analysis, planning, preparation, and conduct of the various facets of a specific campaign"; as quoted from *Military Effectiveness* (London and Boston, 1988), 3 vols., ed. Allan R. Millett and Williamson Murray, vol. 1, p. 12.

4 Edward Mead Earle, ed., *Makers of Modern Strategy* (Princeton, N.J., 1943), p. viii.

5 Basil Henry Liddell Hart, *Strategy*, 2d rev. ed. (New York, 1974), p. 353.

6 Ibid., p. 357 (emphasis added).

7 Ibid., p. 322 (emphasis added).

8 Most thoroughly treated, empathetically by Brian Bond, *Liddell Hart: A Study of His Military Thought* (London, 1977), and critically by John J. Mearsheimer, *Liddell Hart and the Weight of History* (Ithaca, N.Y., 1989).

9 Most fully articulated in Liddell Hart's *The British Way in Warfare* (London, 1932), especially chap. 1, "The Historical Strategy of Britain." See also the observations in Michael Howard, "The British Way in Warfare: A Reappraisal" (Neale Lecture, London, 1975), and in Bond, *Liddell Hart*, chap. 3.

10 Apart from the works cited in notes 8 and 9 above, see Michael Howard,

The Continental Commitment (London, 1972); Correlli Barnett, *Britain and Her Army* (London, 1970); and Paul Kennedy, *The Rise and Fall of British Naval Mastery* (London and New York, 1976) for samples of this criticism.

11 This is probably why Britain and Rome (see Arther Ferrill's essay in this collection) are by far the most frequently studied *historical* examples of grand strategy.

12 Liddell Hart, *Strategy*, p. 353.

13 This has now been dealt with superbly, at least for land warfare, in Martin van Creveld's *Supplying War: Logistics from Wallenstein to Patton* (Cambridge, 1977).

14 For a few examples of later scholarship in this field, see John Brewer, *The Sinews of War* (London and Boston, 1989); P. G. M. Dickson, *The Financial Revolution in England* (London, 1967); Paul Kennedy, *The Rise and Fall of British Naval Mastery;* Jon T. Sumida, *In Defense of Naval Supremacy* (London and Boston, 1989); and David French, *British Economic and Strategic Planning, 1905–1915* (London and Boston, 1982).

15 I have tried to summarize this literature in *The Rise and Fall of the Great Powers* (New York, 1987), chaps. 2 and 3; but see also the excellent recent synthesis in Geoffrey Parker, *The Military Revolution* (Cambridge, 1988), chap. 2.

16 Christopher Andrew, *Théophile Delcassé and the Making of the Entente Cordiale* (London, 1968) is best here.

17 See the quotation cited in note 7 above. For a "Clausewitzian" critique of the Vietnam War, see Harry G. Summers, *On Strategy: A Critical Analysis of the Vietnam War* (New York, 1982).

18 No doubt it is theoretically possible for a small nation to develop a grand strategy, but the latter term is generally understood to imply the endeavors of a power with extensive (i.e., not just local) interests and obligations, to reconcile its means and its ends.

19 See Michael Howard's essay in this collection.

20 In the case of Spain, "political" incorporates "dynastic" and "religious"; in the case of the Soviet Union, it incorporates "social" and "ideological."

21 Michael Howard, "Imperial Cycles: Bucks, Bullets and Bust," *The New York Times Book Review*, 10 January 1988.

22 See Eliot A. Cohen's essay in this collection, especially its early section.

23 See *Discriminate Deterrence*, report of the Commission on Integrated Long-Term Strategy, cochaired by Fred Iklé and Albert Wohlstetter (Washington, D.C., 1988).

Chapter Two. Alliance, Encirclement, and Attrition

1 For a general overview of the field, see Henry L. Snyder, "Marlborough and the Reign of Queen Anne: The Status of Current Studies," *British Studies Monitor* 8, no. 3 (1978), pp. 3–15. In the past decade, the most important contribution to the subject is the continuation of A. J. Veenendaal's series *De briefwisseling van Anthonie Heinsius* (The Hague, 1976). The eleventh volume appeared in 1990, carrying the documents from 1702 through 1710. An earlier work, David Francis's *The First Peninsula War* (London, 1975), makes an important contribution to understanding England's participation in the war in Spain and Portugal. Edward Gregg's *Queen Anne*

(London, 1980) shows her importance and her influence. The literature in English on the Allies continues the biographical approach, but adds new perspectives. The most important contributions include Derek McKay, *Prince Eugene of Savoy* (London, 1977); John Spielman, *Leopold I of Austria* (London, 1977); Charles W. Ingrao, *In Quest and Crisis: Emperor Joseph I and the Habsburg Monarchy* (West Lafayette, Ind., 1979); Geoffrey Symcox, *Victor Amadeus II: Absolutism in the Savoy and State* (London, 1978); and Ragnhild Hatton, *George I: Elector and King* (London, 1978). Others include Linda Frey and Marsha Frey, *A Question of Empire: Leopold I and the War of the Spanish Succession, 1701–05* (Columbus, Ohio, 1983) and Andrew Rothstein, *Peter the Great and Marlborough: Politics and Diplomacy in Converging Wars* (London, 1986). Klaus-Ludwig Feckl's *Preussen in Spanischen Erbfolgekrieg* (Frankfurt am Main, 1979) complements my own *England in the War of the Spanish Succession* (New York, 1987) as a study which focuses on one nation's policies and conduct during the entire war, while J. R. Jones, *Britain and the World, 1649–1815* (London, 1980) provides a wide survey of British foreign policy.

2 Winston S. Churchill, *Marlborough: His Life and Times*, 6 vols. (London, 1933–1938), vol. 1, pp. 3–4.

3 William Blackstone, *Commentaries on the Law of England* (Oxford, 1765), vol. 1, p. 151.

4 John Hattendorf, "English Government Machinery and the Conduct of War, 1702–1713," *War and Society* 3, no. 2 (September 1985), pp. 1–22.

5 Studies of English diplomacy in the period 1698–1701 may be found in M. A. Thomson, "Louis XIV and the Origins of the War of the Spanish Succession," in R. S. Hatton and J. S. Bromley, *William III and Louis XIV* (Liverpool, 1968), pp. 140–161; S. B. Baxter, *William III* (New York, 1966), pp. 364–401; Wolfgang Michael, "The Treaties of Partition and the Spanish Succession," in *The Cambridge Modern History* (New York, 1908), vol. 5, pp. 372–400; Sir George Clark, "From the Nine Years' War to the War of the Spanish Succession," in *The New Cambridge Modern History* (Cambridge, 1970), vol. 6, pp. 381–409; J. W. Smit, "The Netherlands and Europe in the Seventeenth and Eighteenth Centuries" and J. R. Jones, "English Attitudes to Europe in the Seventeenth Century," in J. S. Bromley and E. H. Kossman, eds., *Britain and the Netherlands in Europe and Asia* (London, 1968), pp. 13–55; Jones, *Britain and the World*, pp. 149–162.

6 King William III to Antonie Heinsius, 14 December 1700, in F. J. L. Kramer, ed., *Archives ou correspondence inédite de la maison d'Orange-Nassau* (Leiden, 1909), 3d series, vol. 3, p. 296 (hereafter *Archives . . . Orange-Nassau*).

7 Leicestershire Record Office, Finch Papers, Box 4950, unsigned, undated letter [mid-1700].

8 See Godfrey Davies, "The Reduction of the Army after the Peace of Ryswick, 1697," *Journal of the Society for Army Historical Research* 28 (1950), pp. 15–28.

9 King William III to Antonie Heinsius, 19 November 1700, *Archives . . . Orange-Nassau*, p. 242.

10 King William III to Antonie Heinsius, 17 December 1700 and 21 January 1701, *Archives . . . Orange-Nassau*, pp. 305 and 374.

11 L'Hermitage to the States-General, 12 February 1701, British Library, Additional Manuscript 17,677 WW, fol. 141.

12 Charles Hedges to Alexander Stanhope, 21 February 1701, Public Record Office, State Papers 104/69, fol. 132v.

13 Instructions to the duke of Marlborough, 26 February 1701, Public Record Office, State Papers 104/69, fol. 152ff.

14 "The Treaty of Grand Alliance, 1701," in A. Browning, ed., *English Historical Documents, 1660–1714* (London, 1953), vol. 8, p. 873.

15 *Journal of the House of Lords*, vol. 17, p. 6 (31 December 1701).

16 William Aglionby to Charles Hedges, 13 July 1701, Public Record Office, State Papers 94/75.

17 See, for example, H. A. Lloyd, *The Rouen Campaign, 1590–92* (Oxford, 1973), pp. 37, 70; J. E. Neale, *Queen Elizabeth I* (New York, n.d.), pp. 237–238.

18 Clyve Jones, "The Protestant Wind of 1688: Myth and Reality," *European Studies Review* 3 (1973), p. 216.

19 For general studies of this, see M. A. Thomson, "The Safeguarding of the Protestant Succession, 1702–18," in Hatton and Bromley, *William III and Louis XIV*, pp. 237–251; J. P. Kenyon, *Revolution Principles: The Politics of Party, 1689–1720* (Cambridge, 1977); and G. E. Gregg, "The Protestant Succession in International Politics, 1710–16" (Ph.D. thesis, University of London, 1972).

20 Bonet to King Frederick, 21 January 1701, British Library, Additional Manuscript, 30,000E, fols. 6–7.

21 George Stepney to William Blathwayt, 1 June 1701, British Library, Additional Manuscript, 9720, fols. 3–5.

22 Alexander Stanhope to William Blathwayt, 23 September 1701, British Library, Additional Manuscript, 21,489, fol. 51.

23 Ibid.

24 *Journal of the House of Commons*, vol. 13, p. 665 (10 January 1702).

25 Emperor's ratification, 22 March 1702, Public Record Office State Papers 108/131. Dutch ratification, 8 June 1702, Public Record Office, State Papers 108/337.

26 "Her Majesty's Gracious Declaration at Her first sitting in the Privy Council at St. James," 8 March 1702, Public Record Office, Colonial Office Papers, 324/8, fol. 40.

27 Daniel A. Baugh, "British Strategy during the First World War in the Context of Four Centuries: Blue-Water versus Continental Commitment," in Daniel M. Masterson, ed., *Naval History: The Sixth Symposium of the U.S. Naval Academy* (Wilmington, Del., 1987), pp. 87–89.

28 John Hattendorf, "English Grand Strategy and the Blenheim Campaign of 1704," *International History Review* 5, no. 1 (February 1983), pp. 1–19.

29 See Sir Julian Corbett, *Some Principles of Maritime Strategy*, annotated with an introduction by Eric J. Grove (Annapolis, Md., 1988), Appendix: "'The Green Pamphlet'".

30 John Hattendorf, "The Rákòczi Insurrection in English War Policy, 1703–1711," *The Canadian-American Review of Hungarian Studies* 7 (1980), pp. 91–102.

31 See Fred Charles Iklé, *Every War Must End* (New York, 1971), pp. 1–16.

Chapter Three. British Grand Strategy in World War I

1 Michael Howard, *Grand Strategy*, vol. 4, *U.K. Official History of the Second World War, Military Series* (London, 1972), p. 1.

2 Philip Magnus, *Kitchener: Portrait of an Imperialist* (London, 1958), p. 349.

3 Carl von Clausewitz, "Note of 10 July 1827," in *On War*, Michael Howard and Peter Paret, trans. and eds. (Princeton, 1976), p. 69.

4 Gordon A. Craig, "Delbrück: The Military Historian," in Peter Paret, ed., *Makers of Modern Strategy* (Princeton, N.J., 1986), pp. 326–353.

5 Jean de Bloch [Ivan Bloch], *La Guerre future*, 6 vols. (Paris, 1898); translated as *The Future of War* (Boston, 1899).

6 Alfred von Schlieffen, "Der Krieg in der Gegenwart," in *Gesammellte Schriften* (Berlin, 1913) vol. 1, p. 11.

7 See, for example, L. L. Farrar, *The Short-War Illusion: German Policy, Strategy and Domestic Affairs, August–December 1914* (Santa Barbara, Calif., and Oxford, 1973).

8 For the full debate, see Samuel R. Williamson, *The Politics of Grand Strategy: Britain and France Prepare for War, 1904–1914* (Cambridge, Mass., 1969).

9 For the ambitions of the German right wing and to a large extent the German government, see Fritz Fischer, *Griff nach der Weltmacht* (Düsseldorf, 1962); translated as *Germany's Aims in the First World War* (London, 1967).

10 For the September Program, which envisaged substantial German conquests in Eastern and Western Europe as well as imperial annexations overseas, see Fischer, *Griff nach der Weltmacht*.

11 See, for example, Elizabeth Monroe, *Britain's Moment in the Middle East, 1914–1971* (London, 1981).

12 David French, *British Strategy and War Aims, 1914–1916* (London, 1986). See also his article "The Meaning of Attrition, 1914–1916," in *The English Historical Review* 103, no. 407 (April 1988), p. 385.

13 The best account of the Gallipoli campaign is Robert Rhodes James, *Gallipoli* (London, 1965). The discussions leading to its implementation are documented in Martin S. Gilbert, *Winston Churchill*, vol. 3, *Companion Volume*, part 1 (London, 1972), pp. 278ff.

14 For a comprehensive account of the development of Britain's war effort, see Trevor Wilson, *The Myriad Faces of War: Britain and the Great War, 1914–1918* (Cambridge and Oxford, 1986).

15 See especially Sir William Robertson, *Soldiers and Statesmen*, 2 vols. (London, 1926) and David Lloyd George, *War Memoirs*, 2 vols. (London, 1938).

16 See R. J. Q. Adams and P. J. Poirer, *The Conscription Controversy in Britain, 1900–1918* (London, 1987) pp. 144–195; and French, *British Strategy*, pp. 158–195.

17 V. H. Rothwell, *British War Aims and Peace Diplomacy* (London, 1971), pp. 221–228. Rothwell shows, however, the divisions within the British government over this issue.

18 For a definitive analysis, see J. M. Winter, *The Great War and the British People* (London, 1985).

19 The fullest account of the formulation of British strategy between the wars is in N. H. Gibbs, *Grand Strategy*, vol. 1, *U.K. Official History of the Second World War, Military Series* (London, 1976).

Chapter Four. Churchill and Coalition Strategy in World War II

The author would like to thank William Fuller, Bradford Lee, Williamson Murray, Thomas Schwartz, and Brian Sullivan for their comments on earlier drafts of this article.

1 Bernard Fergusson, ed., *The Business of War: The War Narrative of Major-General Sir John Kennedy* (New York, 1958), p. 115.

2 See Ronald Lewin's masterly *Churchill as Warlord* (1973; repr. New York, 1982), for example, or the memoirs of the secretary to the British Chiefs of Staff, Hastings L. Ismay, *The Memoirs of General Lord Ismay* (New York, 1960), esp. pp. 281–282 and 317–318.

3 See Lewin, *Churchill as Warlord*, pp. 11–12, 266; Ismay, *Memoirs*, pp. 164–166. Thus, for example, Churchill's insistence that none of his orders be taken as such unless they appeared in writing. Even more important was his reliance on the War Cabinet and its subcommittees such as the Defence Committee (Operations) for decision making, and his retention and development of a high-quality secretariat for the War Cabinet.

4 John Colville, *The Fringes of Power: Ten Downing Street Diaries, 1939–1955* (New York, 1985), p. 767.

5 Winston S. Churchill, *The World Crisis*, 4 vols. (New York, 1923–1927), vol. 2, p. 6. Cf. Carl von Clausewitz, *On War*, edited and translated by Michael Howard and Peter Paret (Princeton, N.J., 1976), pp. 605–610.

6 Churchill, *World Crisis*, vol. 2, pp. 217, 349.

7 Ibid., vol. 3, pp. 71–73. "The true and indeed the only attainable political objectives open to Germany in 1916 were the final overthrow of Russia and the winning of Roumania to the side of the Central Empires. . . . The opportunity of General von Falkenhayn, Chief of the German General Staff, was to pronounce the word ROUMANIA. He pronounced instead the word VERDUN." Ibid., vol. 4, pp. 52–54.

8 Clausewitz, *On War*, p. 75. On p. 607 he explains the connection between this strategic holism and the primacy of politics. Cf. Churchill's discussion of the interrelationship between design, tactics, and strategy in *World Crisis*, vol. 1, p. 130.

9 Churchill, *World Crisis*, vol. 2, p. 6.

10 Ibid., vol. 3, p. 14. A penetrating discussion of Churchill's vision of national character (and much else) is in Isaiah Berlin's superb essay *Mr. Churchill in 1940* (Boston, 1949), pp. 12–13, 17–18, 24.

11 Churchill, *World Crisis*, vol. 3, p. 244.

12 Ibid., p. 172. See, too, pp. 164–166 for Churchill's views in the spring of 1918 on British strategy in the event of a German breakthrough.

13 Winston S. Churchill, *Marlborough: His Life and Times*, 6 vols. (New York, 1933–1938). J. H. Plumb finds *Marlborough* a "splendid work of literary art" rather than solid history. This may be the case, but the work is hardly to be dismissed on that account, for in large part it should be read as nothing other than a prolonged and profound meditation on the nature of

statesmanship. Plumb, "The Historian," in A. J. P. Taylor, ed., *Churchill Revised: A Critical Assessment* (New York, 1969), p. 151. Cf. Berlin, *Mr. Churchill in 1940*, p. 31.

14 Churchill, *Marlborough*, vol. 5, p. 9.

15 Ibid., p. 246.

16 Ibid., pp. 553–554.

17 See, for example, ibid., vol. 1, p. 75.

18 Ibid., vol. 3, pp. 133, 353.

19 Compare, for example, Churchill's strategic appreciation in December 1941 with similar efforts in World War I and with descriptions of the problems faced by William and then Marlborough. Warren F. Kimball, ed., *Churchill and Roosevelt: The Complete Correspondence*, 3 vols. (Princeton, N.J., 1984), pp. 294–308. Churchill, *The World Crisis*, vol. 4, pp. 112–122. Churchill, *Marlborough*, vol. 1, pp. 67–83; vol. 2, pp. 233–256; vol. 3, pp. 94–116.

20 Speech to the House of Commons, 8 June 1943. Robert Rhodes James, ed., *Winston Churchill: His Complete Speeches, 1897–1963*, 8 vols. (London, 1974), vol. 7, p. 6789.

21 Ibid., vol. 6, p. 6584 (emphasis added).

22 Winston S. Churchill, *The Second World War*, 6 vols. (Boston, 1948–1953), vol. 3, *The Grand Alliance*, pp. 607–608. That Churchill actually took this view at the time is supported by contemporary accounts and records. See Randolph S. Churchill and Martin Gilbert, *Winston S. Churchill*, 8 vols. (Boston, 1966–1988), vol. 6, *Finest Hour, 1939–1941*, p. 1274; vol. 7, *Road to Victory, 1941–1945*, p. 7.

23 Thus, Alan Brooke confided to his diary in March 1942 his "growing conviction that we are going to lose the war." Arthur Bryant, *The Turn of the Tide* (New York, 1957), p. 276.

24 David Reynolds criticizes the stereotypical view of Churchill as "heroic but uncomplicated," but argues that the decision to fight on in 1940 was driven chiefly by mistaken judgments about the German war economy. This may in some measure be true, but Reynolds underestimates Churchill's single-minded concentration on getting the United States into the war, and his determination (after June 1940, at least) not to yield to Hitler. Reynolds, "Churchill and the British 'Decision' to Fight on in 1940: Right Policy, Wrong Reasons," in Richard Langhorne, ed., *Diplomacy and Intelligence during the Second World War* (Cambridge, 1985), pp. 147–167.

25 Public Record Office, Cabinet Papers (hereafter CAB) 65/8, W.M. (40) 227, War Cabinet meeting of 14 August 1940. On the bases-for-destroyers deal more generally, see J. R. M. Butler, *Grand Strategy*, vol. 2, *U.K. Official History of the Second World War* (London, 1957), pp. 239–246.

26 CAB 65/14, W.M. (40) 232, War Cabinet meeting of 22 August 1940.

27 See, for example, his discussion of the Greek intervention, CAB 65/22, W.M. (41) 44, War Cabinet meeting of 28 April 1941; or the treatment of Far East problems by the cabinet's Defence Committee (Operations), CAB 69/2 D.O. (41) 12, meeting of 9 April 1941.

28 Lewin, *Churchill as Warlord*, p. 110, discusses the importance of Churchill's study of the American Civil War. Churchill's own judgment was that no one in Europe "understood the strength of Abraham Lincoln or the

resources of the United States," in "the noblest and least avoidable of all the great mass-conflicts of which till then there was record." *A History of the English-Speaking Peoples*, 4 vols. (New York, 1956–1958), vol. 4, *The Great Democracies*, pp. 182, 263.

29 On their assessment of their own capabilities, see Fergusson, *Business of War*, p. 198.

30 Bryant, *Turn of the Tide*, p. 233.

31 A useful picture of British official views can be found in Alex Danchev, *Very Special Relationship* (London, 1986), pp. 39–42.

32 H. Duncan Hall, *North American Supply* (London, 1955), p. 420.

33 Ibid., p. 421.

34 W. M. Hancock and M. M. Gowing, *British War Economy* (London, 1949), pp. 367–368. The disparity reflected the superior productivity of American industry, but also the absence in the United States of the disruptions, operational pressures, and sheer fatigue that afflicted a British economy directly exposed to enemy action over six years.

35 Hancock and Gowing, *British War Economy*, p. 373.

36 M. M. Postan, *British War Production* (London, 1952), p. 247.

37 Hancock and Gowing, *British War Economy*, p. 367.

38 In 1943 United States Eighth Air Force dropped between a third and a half as much bomb tonnage on the enemy as did British Bomber Command; in 1944 and 1945 the two air forces dropped approximately equal tonnages. Charles Webster and Noble Frankland, *The Strategic Air Offensive against Germany*, 4 vols. (London, 1961), vol. 4, pp. 454–457.

39 See, for example, Albert C. Wedemeyer, *Wedemeyer Reports* (New York, 1958), pp. 132–136. Wedemeyer's view of Churchill—a "pseudo strategist"—was particularly caustic.

40 William Roger Louis, *Imperialism at Bay: The United States and the Decolonization of the British Empire, 1941–1945* (New York, 1978); Christopher Thorne, *Allies of a Kind: The United States, Britain, and the War against Japan, 1941–1945* (New York, 1978). See some of Roosevelt's cables to Churchill in Kimball, *Churchill and Roosevelt*, vol. 1, pp. 344–345, 357–358, 402–404, 446–447 (two particularly sharp discussions of Britain's role in India); vol. 2, pp. 733–734, 754–755; vol. 3, pp. 3–10, 197ff., 418–421, 424–425, 427.

41 Witness American support for Great Britain during the Falklands War of 1982, for example, or the well-known intelligence cooperation between the two countries which continues to the present day. See Henry A. Kissinger, *The White House Years* (Boston, 1979), p. 90.

42 Colville, *Fringes of Power*, p. 624.

43 Robert E. Sherwood, *Roosevelt and Hopkins: An Intimate History* (New York, 1948), pp. 363–364.

44 Ismay, *Memoirs*, pp. 258–259.

45 *Speeches*, vol. 4, p. 6238 (emphasis added).

46 "Churchill's speech to the Congress was so informative that Congressmen were louder than ever in their complaints that, 'The only time we get to find out what's going on in the war is when the British Prime Minister visits Washington and tells us.'" Sherwood, *Roosevelt and Hopkins*, p. 729.

47 The correspondence has been reproduced by the Soviets, in translation. Ministry of Foreign Affairs of the USSR, *Stalin's Correspondence with Churchill and Attlee, 1941–1945* (New York, 1965). This collection contains some five hundred messages passed in either direction, many of them brief or purely formal. On the other hand, there is some good humor in it—for example, Churchill's remarks to Stalin about not letting De Gaulle see the films *Kutuzov* or *Lady Hamilton*. Churchill to Stalin, 19 December 1944, p. 287.

48 Broadcast, 1 October 1939, in *Speeches*, vol. 6, p. 6161.

49 Broadcast, 22 June 1941, ibid., pp. 6429–6431. See, too, Colville, *Fringes of Power*, p. 404.

50 See Ismay, *Memoirs*, p. 235. The ebb and flow of British confidence in the Red Army from the spring of 1941 through the beginning of 1943 are traced in F. H. Hinsley et al., *British Intelligence in the Second World War*, 4 vols. (London, 1979–1988), vol. 1, pp. 429–483; vol. 2, pp. 67–128.

51 J. M. Gwyer, *Grand Strategy*, vol. 3, part 1, *U.K. Official History of the Second World War* (London, 1964), p. 98.

52 As argued in an inaccurate and tendentious article by Martin Kitchen, "Winston Churchill and the Soviet Union during the Second World War," *Historical Journal* 30, no. 2 (June 1987), pp. 415–436. Cf. Hancock and Gowing, *British War Economy*, pp. 359–365; Michael Howard, *Grand Strategy*, vol. 4, *U.K. Official History of the Second World War* (London, 1970), pp. 31–47.

53 CAB 69/4, D.O. (42) 15, Defence Committee (Operations) meeting of 13 July 1942.

54 CAB 65/30, W.M. (42) 64, War Cabinet meeting of 18 May 1942.

55 Gilbert, *Churchill*, vol. 7, p. 34.

56 See CAB 69/2, D.O. (41) 64, Defence Committee (Operations) meeting of 15 October 1941.

57 Soviet Ministry of Foreign Affairs, *Stalin's Correspondence*, p. 13.

58 Quoted in Howard, *Grand Strategy*, vol. 4, p. 369.

59 See, for example, his critique of a Chiefs of Staff strategy paper in CAB 59/4, D.O. (42) 17, Defence Committee (Operations) meeting of 16 November 1942.

60 CAB 65/29, W.M. (42) 17, War Cabinet meeting of 6 February 1942.

61 Gilbert, *Churchill*, vol. 7, p. 239.

62 Ibid., p. 373.

63 Kimball, *Churchill and Roosevelt*, vol. 2, pp. 389–402; Gilbert, *Churchill*, vol. 7, pp. 384–385.

64 See Gilbert, *Churchill*, vol. 7, pp. 923–929.

65 Speech to the House of Commons, 22 February 1944, in *Speeches*, vol. 7, p. 6887.

66 See, for example, his speeches in the United States on 25 and 31 March 1949, in *Speeches*, vol. 7, pp. 7795–7811.

67 Colville, *Fringes of Power*, p. 658. Diary entry, 1 January 1953. Churchill loathed and feared communism, which he thought every bit as wicked as nazism. However, he thought communism rested on so false a conception of human nature and society that it could not last. Nazism, on the other

hand, precisely because it appealed—in depraved or perverted form—to more durable and bellicose human instincts such as the desire for domination and glory, posed the greater menace. See Winston S. Churchill, *The Aftermath* (New York, 1929), pp. 60–66; idem, "Mass Effects and Modern Life," in *Amid These Storms: Thoughts and Adventures* (New York, 1932), pp. 255–268; idem, "Hitler and His Choice," in *Great Contemporaries* (London, 1937), pp. 261–272.

68 See H. Duncan Hall and C. C. Wrigley, *Studies of Overseas Supply* (London, 1956), pp. 46–65. Great Britain herself produced approximately 70 percent, most of the balance coming from the United States. Hall, *North American Supply*, p. 428. Canadian production included more than 16,000 aircraft, over a quarter of which were for British use.

69 CAB 65/14, W.M. (40) 217, War Cabinet meeting of 1 August 1940. This is in addition to the Dominion subjects who enlisted as individuals in British outfits.

70 Gwyer, *Grand Strategy*, vol. 3, part 1, pp. 223–230.

71 On Australian-British relations in December 1941, see Butler, *Grand Strategy*, vol. 2, part 2 (London, 1964), pp. 408–409. More generally, see D. M. Horner, *High Command: Australia and Allied Strategy, 1939–1945* (Canberra, 1982).

72 On the less problematic relations between Great Britain and Canada, see C. P. Stacey, *Arms, Men and Governments: The War Policies of Canada, 1939–1945* (Ottowa, 1970), pp. 137–251. The Canadian prime minister, Mackenzie King, insisted that the British government secure Canadian approval before sending any Canadian forces to intervene in the Greek civil war.

73 See the exchanges in Kimball, *Churchill and Roosevelt*, vol. 1, pp. 372–379, 402–404, 446–450. On the other hand, it should be noted that Roosevelt sided with Churchill in some cases, notably in pressuring the Australians to keep their Ninth Division in the Middle East in October 1942. Ibid., p. 644.

74 CAB 65/23, W.M. (41) 94, War Cabinet meeting of 18 September 1941.

75 CAB 65/23, W.M. (41) 98, War Cabinet meeting of 29 September 1941.

76 See Butler, *Grand Strategy*, vol. 2, pp. 262–263.

77 See J. M. Lee, *The Churchill Coalition, 1940–1945* (Hamden, Conn., 1980), pp. 32, 142, 159, 164.

78 *Speeches*, vol. 6, p. 6565.

79 Stacey, *Arms, Men, and Governments*, p. 154.

80 The story of this relationship is well told in François Kersuady, *Churchill and De Gaulle* (New York, 1983).

81 See Elisabeth Barker, *Churchill and Eden at War* (London, 1978), pp. 41ff.

82 See Berlin, *Mr. Churchill in 1940*, pp. 12–13, 17–18.

83 Butler, *Grand Strategy*, vol. 2, pp. 263–264; Colville, *Fringes of Power*, p. 232.

84 See Gilbert, *Churchill*, vol. 7, pp. 1008ff.

85 CAB 69/6, D.O. (44) 6, Defence Committee (Operations) meeting of 13 April 1944.

86 See John Ehrman, *Grand Strategy*, vol. 5, *August 1943–September 1944* (London, 1956), pp. 297–304.

87 *The Second World War*, vol. 1, p. 547.

88 CAB 65/7, W.M. (40) 177, War Cabinet meeting of 23 June 1940.

89 Churchill's support for the intervention in Iraq despite vehement opposition by General Archibald Wavell, the commander in chief on the scene, can be traced in CAB 69/2, D.O. (41) 24, 25, and 26, which cover the Defence Committee (Operations) meetings of 3, 6, and 8 May 1941.

90 See Howard, *Grand Strategy*, vol. 4, pp. 230–231. Churchill pointed out many months before (in November 1942) that "when a nation is thoroughly beaten in war it does all sorts of things which no one can imagine beforehand." Ibid., p. 231.

91 Ibid., pp. 498–499.

92 Ibid., pp. 486, 523.

93 Kimball, *Churchill and Roosevelt*, vol. 2, pp. 348–351, 369.

94 Percy Schramm, ed., *Kriegstagebuch des Oberkommandos der Wehrmacht, 1940–1945* (Munich, 1982), vol. 3, pp. 736, 1161.

95 See Walter Warlimont, *Inside Hitler's Headquarters*, trans. R. H. Barry (New York, 1964), pp. 381ff.

96 Churchill, *Marlborough*, vol. 3, p. 315.

97 See Berlin, *Mr. Churchill in 1940*, pp. 15–17. For Churchill's own view, see "Consistency in Politics," *Amid These Storms*, pp. 39–47. Colville, *Fringes of Power*, p. 127, notes that Churchill's rhetoric often made him seem impetuous, but that in fact he rarely acted "without careful consideration."

98 For criticism of this approach, see Barry Watts, *The Foundations of U.S. Air Doctrine: The Problem of Friction in War* (Maxwell Air Force Base, 1984); Edward N. Luttwak, *Strategy: The Logic of War and Peace* (Cambridge, Mass., 1987).

99 Howard, *Grand Strategy*, vol. 4, p. 295.

100 CAB 69/2, D.O. (41) 4, Defence Committee (Operations) meeting of 13 January 1941. He said this with reference to plans advanced by the Royal Air Force to bomb Germany as a path to victory.

101 Winston S. Churchill, *A Roving Commission* (1930; repr. New York, 1951), p. 232.

102 CAB 69/1, D.O. (40) 39, Defence Committee (Operations) meeting of 31 October 1940.

103 Speech of 19 May 1943, in *Speeches*, vol. 7, p. 6783.

104 The discussion here draws in large part on my article "Churchill at War," *Commentary* 83, no. 5 (May 1987), pp. 40–49.

105 See his remarkable essay "Painting as a Pastime," in *Amid These Storms*, pp. 305–320.

106 Ibid., p. 309.

107 Ibid., pp. 310–311.

108 Churchill, *Marlborough*, vol. 1, p. 94. Churchill described Halifax as "the foremost statesman of these times."

109 Churchill, *My Early Life*, p. 331.

110 See the discussion in Clausewitz, *On War*, pp. 148–150.

111 John Ehrman, *Grand Strategy*, vol. 6, *U.K. Official History of the Second World War* (London, 1956), p. 333. Ehrman's summative assessment of

Churchill's role in the war (pp. 333–338) is the best short statement on the subject.

Chapter Five. The Grand Strategy of the Roman Empire

1 Gibbon, *The History of the Decline and Fall of the Roman Empire* with introduction, notes, and appendices by J. B. Bury (London, 1909–1914), vol. 1, p. 1.

2 See, for example, Jozef Wolski, "Le Rôle et l'importance des guerres de deux fronts dans la décadence de l'Empire romain," *Klio* 62 (1980), pp. 411–423. See also D. Haupt and H. G. Horn, eds., *Studien zu den Militärgrenzen Roms* (Cologne, 1977).

3 Luttwak, *The Grand Strategy of the Roman Empire: From the First Century A.D. to the Third* (Baltimore, 1976).

4 See the comments of F. Millar, "Emperors, Frontiers, and Foreign Relations, 31 BC–AD 378," *Britannia* 13 (1982), pp. 1–23; and of J. C. Mann, "The Frontiers of the Principate," in H. Temporini, ed., *Aufstieg und Niedergang der römischen Welt* (Berlin and New York, 1972), 2d series, vol. 1, pp. 508–533.

5 See Luttwak, *Grand Strategy*, pp. 74–75.

6 In fact, Luttwak carefully qualifies his argument and states it much less rigidly than it has been paraphrased by others and even by me, here, and in my book *The Fall of the Roman Empire: The Military Explanation* (London and New York, 1986), pp. 23–50. I am indebted to Thames and Hudson Ltd. for permission to use parts of my book in this essay.

7 On the third century, see Michael Grant, *The Climax of Rome* (London, 1968) and Ramsay MacMullen, *Roman Government's Response to Crisis, A.D. 235–337* (New Haven, 1976).

8 There has been considerable disagreement about whether the new grand strategy was introduced by Diocletian or Constantine. Luttwak actually argues for Diocletian. See my discussion in *Fall*, p. 43 and n. 64.

9 Lawrence Keppie, *The Making of the Roman Army* (Totowa, N.J., 1984), pp. 191–198, presents the development of Roman frontier defense in the manner referred to above and avoids the expression "grand strategy." See also F. Millar, "Emperors," p. 2.

10 Luttwak, *Grand Strategy*, p. 111.

11 F. Millar, "Emperors," p. 6: "If the Emperor possessed any secretarial staff specifically for the conduct of frontier policy or diplomacy, all trace of it has disappeared."

12 See Keppie, *Roman Army*, pp. 77–78. It should be noted that standing armies need not be professional, i.e., made up of long-term volunteers. A standing army may be raised by conscription and maintained by frequently changing short-term recruits or draftees. See also John Wacher, *The Roman Empire* (London, 1987), p. 17, and Stephen Dyson, *The Creation of the Roman Frontier* (Princeton, N.J., 1985).

13 For a recent study with full documentation and bibliography, see Kurt Raaflaub, "Die Militärreformen des Augustus und die politische Problematik des frühen Prinzipats," *Saeculum Augustum* 1 (Darmstadt, 1987), pp. 246–307.

14 Augustus, *Res gestae*, 25–33. For a recent, good translation in English, based on the edition edited by P. A. Brunt and J. M. Moore, see David C.

Braund, *Augustus to Nero: A Sourcebook on Roman History, 31 BC–AD 68* (Totowa, N.J., 1985), pp. 14–22.

15 Colin Wells, *The German Policy of Augustus* (Oxford, 1972), notwithstanding the ingenious arguments of Josiah Ober, "Tiberius and the Political Testament of Augustus," *Historia* 31 (1982), pp. 306–328.

16 *Res gestae*, 26 and 30.

17 Luttwak recognizes this basic point, but his emphasis is on the "decreased elasticity" of the second century defense system which he says became "dangerously thin." See *Grand Strategy*, p. 126.

18 For example, three excellent studies are Graham Webster, *The Roman Imperial Army* (London, 1969), G. R. Watson, *The Roman Soldier* (London, 1969), and Michael Speidel, *Roman Army Studies* (Amsterdam, 1984).

19 For a discussion of the tactical formations of the Roman army, see Everett S. Wheeler, "The Legion as Phalanx," *Chiron* 9 (1979), pp. 303–318, which also contains bibliographical references to most of the modern literature on the topic.

20 (Harrisburg, Pa., 1987), pp. 8–299.

21 Ibid., pp. 136–137.

22 (New York, 1950), pp. 49–50.

23 Josephus, *The Jewish War* 3.10.2, trans. William Whiston.

24 On this point Luttwak is inconsistent and in fact contradictory. On p. 1 he denies that Roman strength "derived from a tactical superiority on the battlefield," but on p. 42 he stresses Roman "tactical superiority." See the review by Zvi Yavetz in *The New Republic*, 21 May 1977, pp. 55–57, esp. p. 57.

25 There has been an extensive debate on the importance of the gold of Dacia, but whether it was significant under Trajan or not, the long-range effect was minimal. Indeed, Dacia proved the most difficult province to defend, and it was the first to be permanently abandoned. For the debate on Dacian gold, see Lino Rossi, *Trajan's Column and the Dacian Wars* (London, 1971). On the economics of war during the Roman Republic, see W. V. Harris, *War and Imperialism in Republican Rome, 327–70 B.C.* (Oxford, 1979).

26 Chester G. Starr, *The Roman Empire, 27 B.C.–A.D. 476* (Oxford, 1982), pp. 86–89, and Keith Hopkins, *Journal of Roman Studies* 70 (1980), pp. 101–125.

27 See Luttwak, *Grand Strategy*, pp. 13–17.

28 See Brent Shaw, "Fear and Loathing: The Nomad Menace and Roman Africa," in C. M. Wells, ed., *Roman Africa: The Vanier Lectures, 1980* (Ottawa, 1982), pp. 29–50.

29 For the eastern frontier, see F. Millar, *The Roman Empire and Its Neighbors* (New York, 1965); M. A. R. Colledge, *The Parthians* (New York, 1967); I. Browning, *Palmyra* (London, 1979); and Freya Stark, *Rome on the Euphrates: The Story of a Frontier* (New York, 1966).

30 The best book is E. Mary Smallwood, *The Jews in the Roman World* (Leiden, 1976), which is heavily documented.

31 See Starr, *Roman Empire*, p. 121.

32 E. A. Thompson, "Early Germanic Warfare," *Past and Present* 14 (1958), p. 18.

33 Ferrill, *Fall*, p. 60.

34 Stewart I. Oost, *Galla Placidia Augusta* (Chicago, 1968), pp. 96–97, n. 32. See also D. Jones, "The Sack of Rome," *History Today* 20 (1970), pp. 603–609.

35 See the interesting comments, and some reservations, of Ramsay Mac-Mullen, "Notes on Romanization," *Bulletin of the American Society of Papyrologists* 21 (1984), pp. 161–177.

36 G. L. Cheesman, *The Auxilia of the Roman Imperial Army* (Oxford, 1914); D. B. Saddington, *The Development of the Roman Auxiliary Forces from Caesar to Vespasian (40 B.C.–A.D. 79)* (Hare, 1982); and P. A. Holder, *Studies in the Auxilia of the Roman Army from Augustus to Trajan* (Oxford, 1980).

37 A classic study is J. H. Oliver's edition of Aelius Aristides, *On Rome*, in *Transactions of the American Philosophical Society* n.s. 43, no. 4 (1953).

38 The basic study is G. Forni, *Il reclutamento delle legione da Augusto a Diocleziano* (Milan, 1953); see also P. A. Brunt, "Conscription and Volunteering in the Roman Imperial Army," *Scripta Classica Israelica* 1 (1974), pp. 90–115; and J. C. Mann, *Legionary Recruitment and Veteran Settlement during the Principate* (London, 1983).

39 Ferrill, *Fall*, p. 68 (see also pp. 43, 56, and 153).

40 Keppie, *Roman Army*, pp. 196–197.

41 A. H. Jones, *Augustus* (London, 1970), p. 74.

42 Lukas De Blois, *The Policy of the Emperor Gallienus* (Leiden, 1976), pp. 29–30. See also John Eadie, "The Development of Roman Mailed Cavalry," *Journal of Roman Studies* 57 (1967), pp. 161–173.

43 See my discussion of this point in *The Fall of the Roman Empire*, pp. 41–43.

44 Zosimus, *Historia nova* 2.34, trans. James J. Buchanan and Harold T. Davis (San Antonio, Tex., 1967).

45 Gibbon, *Decline and Fall*, vol. 2, pp. 188–189. On the size of the mobile army, see Dietrich Hoffmann, *Das spätromische Bewegungsheer und die Notitia Dignitatum* (Düsseldorf, 1969), vol. 1, p. 304. See also Ramsay MacMullen, "How Big Was the Roman Army?" *Klio* 62 (1980), pp. 451–460. One of the best books on all matters affecting the late Roman Empire is Emilienne Demougeot, *La Formation de l'Europe et les invasions barbares*, vol. 2, *De l'Avènement de Dioclétien (284) à l'occupation germanique de l'empire romain d'Occident* (Paris, 1979). In English, in addition to Gibbon, the best books are J. B. Bury, *History of the Later Roman Empire*, 2 vols. (1923; repr. London and New York, 1957) and A. H. M. Jones, *The Later Roman Empire, 284–602*, 4 vols. (Oxford, 1964). See also Ramsay MacMullen, *Corruption and the Decline of Rome* (New Haven, 1988), which appeared only after this essay was written.

46 T. Mommsen, "Das römische Militärwesen seit Diocletian," *Hermes* 24 (1889), pp. 195–275.

47 Starr, *Roman Empire*, p. 123.

48 Ferrill, *Fall*, pp. 46–47.

49 See W. Goffart, "The Date and Purpose of Vegetius' *De re militari*," *Traditio* 33 (1977), pp. 65–100; and Ferrill, *Fall*, p. 179, n. 196.

50 Quoted in Goffart, "Date and Purpose," pp. 82–83.

51 *De re militari* 1.20 (translation mine).

52 On the above points generally, see my *Fall*. See also Martin Sieff, "When the Grunts Fall, So Do Their Empires," *The Washington Times*, 24 March 1987, pp. D1–2.

Chapter Six. Managing Decline

1 Peter Jenkins, "Patient Britain," *The New Republic*, 23 December 1985, p. 15.

2 J. H. Elliott, *Imperial Spain, 1469–1716* (London, 1963), p. 378.

3 This lecture was given before the publication of Paul Kennedy's *The Rise and Fall of the Great Powers* (New York, 1987).

4 See my *The Count-Duke of Olivares: The Statesman in an Age of Decline* (New Haven, 1986). The pages that follow are based heavily on this book, to which readers are referred for the necessary documentation, and additional detail.

5 For the Spanish army in the Netherlands, its difficulties and achievements, see Geoffrey Parker, *The Army of Flanders and the Spanish Road, 1567–1659* (Cambridge, 1972).

6 For the Pax Hispanica, see H. R. Trevor-Roper, "Spain and Europe, 1598–1621", *The New Cambridge Modern History*, vol. 4 (Cambridge, 1970), chap. 9.

7 For the *arbitristas*, see my "Self-perception and Decline in Early Seventeenth-Century Spain," *Past and Present* 74 (1977), pp. 41–61, and the sources there cited.

8 John H. Elliott and José F. de la Peña, *Memoriales y cartas del Conde Duque de Olivares*, 2 vols. (Madrid, 1978–1980), vol. 1, p. 52.

9 Ibid., vol. 1, p. 98.

10 Speech as quoted by Juan Yañez, *Memorias para la historia de don Felipe III* (Madrid, 1723), p. 117.

11 Cited in Elliott, *Count-Duke*, p. 74.

12 Ibid., p. 78.

13 Elliott and La Peña, *Memoriales y cartas*, vol. 1, p. 98.

14 Cited in J. H. Elliott, *Richelieu and Olivares* (Cambridge, 1984), p. 77.

15 Ibid., p. 85.

16 Cited in Elliott, *Count-Duke*, p. 231.

17 Memorandum by Olivares (February 1635), cited in John H. Elliott, *El Conde-Duque de Olivares y la herencia de Felipe II* (Valladolid, 1977), 91.

18 Elliott, *Richelieu and Olivares*, p. 87.

19 C. R. Boxer, *Salvador de Sá and the Struggle for Brazil and Angola, 1602–1686* (London, 1952), p. 60.

20 J. H. Elliott, "A Question of Reputation? Spanish Foreign Policy in the Seventeenth Century," *The Journal of Modern History* 55 (1983), pp. 475–483.

21 Elliott, *Count-Duke*, p. 491.

22 Ibid., p. 495.

23 Cited in J. H. Elliott, *The Revolt of the Catalans* (Cambridge, 1963), p. 523.

24 Cited in Elliott, *Richelieu and Olivares*, p. 153.

25 Cited in ibid., p. 159.

26 Elliott, *Count-Duke*, p. 534.

27 Giovanni Botero, *The Reason of State*, trans. P. J. and D. P. Waley (London, 1956), pp. 9–12.

28 Elliott, *Count-Duke*, p. 464.

Chapter Seven. Total War for Limited Objectives

1 See, inter alia, Manfred Messerschmidt, "Preussens Militär in seinem Gesellschaftlichen Umfeld," in *Preussen im Rückblick*, ed. H. J. Pühle and H.-U. Wehler, *Geschichte und Gesellschaft*, special number 6 (1980), pp. 43–88; Michael Geyer, "Professionals and Junkers: German Rearmament and Politics in the Weimar Republic," in *Social Change and Political Development in Weimar Germany*, ed. R. Bessel and E. J. Feuchtwanger (London, 1981), pp. 77–133; Klaus-Jürgen Müller, *Armee und Drittes Reich 1933–1939* (Paderborn, 1987).

2 See Holger Herwig, "The Dynamics of Necessity: German Military Policy during the First World War," in *Military Effetiveness*, vol. 1, *The First World War*, ed. A. R. Millett and W. Murray (Boston, 1988), pp. 80–115; Timothy T. Lupfer, "The Dynamics of Doctrine: The Changes in German Tactical Doctrine during the First World War," *Leavenworth Papers* 4 (Ft. Leavenworth, Kan., 1981); and Bruce Gudmundsson's forthcoming monograph *The Forlorn Hope: Tactical Innovation in the German Army, 1914– 1918*.

3 Martin van Creveld, *Fighting Power* (Westport, Conn., 1982); Max Hastings, *Overlord: D-Day, June 6, 1944* (New York, 1984); Trevor N. Dupuy, *Numbers, Prediction and War* (New York, 1979).

4 See, inter alia, Williamson Murray, "JCS Reform: A German Example?" in *JCS Reform: Proceedings of the Conference*, ed. S. T. Ross (Newport, R.I., 1985), pp. 80–93; Holger Herwig, "From Tirpitz Plan to Schlieffen Plan: Some Observations on German Military Planning," *Journal of Strategic Studies* 9 (1986), pp. 53–63; and Martin van Creveld, "On Learning from the Wehrmacht and Other Things," *Military Review* 68 (January 1988), pp. 62–71.

5 See Gregor Schöllgen, "Sicherheit durch Expansion? Die Aussen- politischen Lageanalysen der Hohenzollern im 17. and 18. Jahrhundert im Lichte des Kontinuitätsproblems in der Preussischen und Deutschen Geschichte," *Historisches Jahrbuch* 104 (1984), pp. 22–45; Christopher Duffy, *Frederick the Great: A Military Life* (London, 1985); and Werner Gembruch, "Struktur des preussischen Staates und aussenpolitische Situa- tion zu Beginn der Herrschaft Friedrichs des Grossen," in *Friedrich der Grosse und das Militärwesen seiner Zeit: Vorträge zur Militärgeschichte*, vol. 8 (Herford, 1987), pp. 9–32. For the logistics problem, see Martin van Creveld, *Supplying War: Logistics from Wallenstein to Patton* (New York, 1977), pp. 28ff.

6 Dennis E. Showalter, "The Retaming of Bellona: Prussia and the Institu- tionalization of the Napoleonic Legacy, 1815–1871," *Military Affairs* 44 (1980), pp. 57–63. See Carolyn Shapiro, "Napoleon and the Nineteenth- century Concept of Force," *Journal of Strategic Studies* 11 (1988), pp. 509– 519.

7 Carl von Clausewitz, *On War*, edited and translated by Michael Howard and Peter Paret (Princeton, N.J., 1976), pp. 484, 595–596.

8 Dennis E. Showalter, *Railroads and Rifles: Soldiers, Technology and the Unification of Germany* (Hamden, Conn., 1975).

9 On this issue, see the recent major syntheses by Thomas Nipperdey, *Deutsche Geschichte 1800–1866* (Munich, 1983); and Heinrich Lutz, *Zwischen Habsburg und Preussen: Deutschland 1815–1866* (Berlin, 1985).

10 See the general discussion by Thomas Nipperdey, "Der Föderalismus in der deutschen Geschichte," *Bydragen en Medelingen Betreffende de Geschiedenis der Nederlanden* 94 (1979), pp. 497–547; and George Windell, "The Bismarckian Empire as a Federal State, 1866–1870: A Chronicle of Failure," *Central European History* 2 (1969), pp. 291–311. N. M. Hope, *The Alternative to German Unification: The Anti-Prussian Party in Frankfurt, Nassau, and the Two Hessen, 1859–1867* (Wiesbaden, 1973) clearly establishes the depth of anti-Prussian antagonism in one "mediatized" area.

11 For Moltke's views on the relationship of war and politics, see Rudolf Stadelmann,*Moltke und der Staat* (Krefeld, 1950); and Gunther E. Rothenberg's update, "Moltke, Schlieffen, and the Doctrine of Strategic Envelopment," in Peter Paret, ed., *Makers of Modern Strategy* (Princeton, N.J., 1986), pp. 296–325. Eberhard Kessel, *Moltke* (Stuttgart, 1957) remains the best general biography. On the problems posed by the war against the French Republic, see, most recently, Eberhard Kolb, "Der schwierige Weg zum Frieden. Das Problem der Kriegsbeendigung 1870/71," *Historische Zeitschrift* 241 (1985), pp. 51–79.

12 See, inter alia, Paul Schroeder, "World War I as Galloping Gertie: A Reply to Joachim Remak," *Journal of Modern History* 44 (1972), pp. 332–334; and Gregor Schöllgen, *Imperialismus und Gleichgewicht. Deutschland, England und die orientalische Frage 1871–1914* (Munich, 1984).

13 On the links between *Sammlungspolitik* and *Weltpolitik*, see particularly Dirk Stegmann, *Die Erben Bismarcks* (Koln and Berlin, 1970), pp. 63ff.; and V. R. Berghahn, *Der Tirpitz-Plan. Genesis und Verfall einen innenpolitischen Krisenstrategie unter Wilhelm II* (Düsseldorf, 1970).

14 Modris Eksteins, "When Death Was Young . . . : Germany, Modernism, and the Great War," in *Ideas into Politics: Aspects of European History, 1880–1950*, ed. R. J. Bullen et al. (London, 1984), pp. 25–35, is an eloquent brief presentation of Germany in the modern era as "more modern than we have had the courage to admit" (p. 33). See as well the contributions in *Another Germany: A Reconsideration of the Imperial Era*, ed. J. Remak and J. Dukes (Boulder, Colo., 1987).

15 This line of argument is developed in my forthcoming monograph *Tannenberg: Clash of Empires*.

16 See Peter Winzen, *Bülows Weltmachtkonzept. Untersuchungen zur Frühphase seiner Aussenpolitik 1897–1901* (Boppard, 1977); Barbara Vogel, *Deutsche Russlandpolitik. Das Scheitern der deutschen Weltpolitik und Bülow, 1901–1906* (Düsseldorf, 1973); and R. R. Mennig, "The Collapse of 'Global Diplomacy': Germany's Descent into Isolation, 1906–1909" (Ph.D. diss., Brown University, 1986).

17 An excellent case study is A. Moritz, *Das Problem des Präventikrieges in der deutschen Politik während der ersten Marokkokrise* (Bern, 1974).

18 Riezler's prewar views are best expressed in *Die Erforderlichkeit des Unmöglichen* (Munich, 1913). See Wayne C. Thompson, *The Eye of the Storm: Kurt Riezler and the Crisis of Modern Germany* (Iowa City, 1980);

and Andreas Hillgruber, "Riezlers Theorie des kalkulierten Risikos und Bethmann-Hollwegs politische Konzeption in der Julikrise 1914," *Historische Zeitschrift* 202 (1966), pp. 333–351.

19 See Headquarters of the Eighth Army Corps to Oberpräsident der Rheinprovinz concerning measures for arresting political unreliables on the declaration of a state of war, 3 April 1914; with the statement of the General Staff of the Field Army to federal war ministers and corps districts regarding preparation of the *Burgfrieden,* 13 August 1914, in *Militär und Innenpolitik im Weltkrieg 1914–1918,* ed. W. Deist, 2 vols. (Düsseldorf, 1970), vol. 1, pp. 185–187, 193–194.

20 *Immediatvortrag* of 24 October 1908, in Bundesarchiv/Militararchiv (hereafter BA/MA), Record Group RM5/1607.

21 See particularly Ivo N. Lambi, *The Navy and German Power Politics, 1862–1914* (Boston, 1984), esp. pp. 332ff.

22 Stig Förster, "Facing 'People's War': Moltke the Elder and German Military Options after 1871," *Journal of Strategic Studies* 10 (1987), pp. 209–230.

23 See, inter alia, Manfred Rauh, *Das Parlamentarisierung des Deutschen Reiches* (Düsseldorf, 1977); the critique by Dieter Langewiesche, "Das deutsche Kaiserreich—Bemerkungen zur Diskussion über Parlamentarisierung und Demokratisierung Deutschlands," *Archiv für Sozialgeschichte* 19 (1979), pp. 628–642; and Beverly Heckart, *From Bassermann to Bebel: The Grand Bloc's Quest for Reform in the Kaiserreich, 1900–1914* (New Haven, 1974). For the difficulties of establishing a sense of common purpose in the Reichstag, see particularly James J. Sheehan, "Political Leadership in the German Reichstag, 1871–1918," *American Historical Review* 74 (1968), pp. 511–528; and William Clagget et al., "Political Leadership and the Development of Political Cleavages: Imperial Germany, 1871–1912," *American Journal of Political Science* 26 (1982), pp. 643–664.

24 See Dieter Groh, *Negative Integration und revolutionärer Attentismus. Die deutsche Sozialdemokratie am Vorabend des Ersten Weltkrieges* (Frankfurt, 1973); and Kenneth R. Calkins, "The Uses of Utopianism: The Millenarian Dream in Central European Social Democracy before 1914," *Central European History* 15 (1982), pp. 124–148.

25 On this controversial issue, see the comprehensive analysis by Bruno Thoss, "Nationale Rechte, militärische Führung und Diktaturfrage in Deutschland 1913–1923," *Militärgeschichtliche Mitteilungen* 42 (1987), pp. 27–76.

26 The specific nature of German anxieties regarding Russia's military potential can be traced through "Die Militärpolitische Lage Deutschlands," 2 December 1911, in *Kriegsrüstung und Kriegswirtschaft,* ed. Reichsarchiv, *Anlageband* (Berlin, 1930), pp. 126ff.; "Die wichtigste Veränderungen im Heerwesen Russlands im Jahre 1911," BA/MA, RM5/1486, Russland Militärisches, April 1892–April 1914; "Nachrichten über die militärische Lage in Russland," 21 November 1912, in Politischen Archiv des Auswärtigen Amtes (hereafter PAAA), Russland 72/82; and "Die wichtigste veränderungen im Heerwesen Russlands im Jahre 1913," BA/MA, RM5/1486.

27 Michael Geyer, *Deutsche Rüstungspolitik 1860–1980* (Frankfurt, 1984), pp. 83ff., highlights the quantity-versus-quality issue. Cf. Stig Förster, *Der Doppelte Militärismus. Die deutsche Heeresrüstungspolitik zwischen StatusQuo-Sicherung und Aggression 1890–1913* (Stuttgart, 1983).

28 Schlieffen to his sister Marie, 13 November 1892, in Eberhard Kessel, ed., *Generalfeldmarschall Graf Alfred Schlieffen. Briefe* (Göttingen, 1958), pp. 295–298; Prince Krafft zu Hohenloe-Ingelfingen, *Militärische Briefe*, vol. 2, *Ueber Infanterie*, 2d ed. (Berlin, 1886), pp. 19ff.

29 Jack R. Dukes, "Militarism and Arms Policy Revisited: The Origins of the German Army Law of 1913," in *Another Germany: A Reconsideration of the Imperial Era*, ed. Jack R. Dukes and Joachim Remak (Boulder, Colo., 1988), pp. 19–39; and Helmut Altrichter, *Konstitutionalismus und Imperialismus. Der Reichstag und die deutsch-russischen Beziehungen 1890–1914* (Frankfurt, 1977), which incorporates a comprehensive discussion of successive military budgets in the context of their effect on diplomacy.

30 Heinrich Hasenbein, "Die parlamentarische Kontrolle des militärischen Oberbefehls im Deutschen Reich von 1871 bis 1918," (Diss., Göttingen, 1968).

31 Karl von Einem, *Erinnerungen eines Soldaten 1853–1933*, 2d ed. (Leipzig, 1933), p. 107.

32 Erich Ludendoff, *Mein militärischer Werdegang* (Munich, 1933), p. 152.

33 Wilhelm Groener, *Der Weltkrieg und seine Probleme* (Berlin, 1920), p. 51.

34 The evolution of Schlieffen's thought can be traced in the *Aufgaben* for 1903 and 1904 in Generalstab des Heeres, Kriegswissenschaftliche Abteilung (ed.), *Dienstschriften des Chefs des Generalstabes der Armee Generalfeldmarschalls Graf von Schlieffen*, vol. 1, *Die taktisch-strategischen Aufgaben aus den Jahren 1891–1905* (Berlin, 1937), pp. 103ff.; and the Staff Rides East for 1901 and 1903, ibid., vol. 2, *Die Grossen Generalstabsreisen-Ost aus den Jahren 1891–1905* (Berlin, 1937), pp. 222ff. and 300ff. Cf. Lothar Burchardt, "Operatives Denken und Planen von Schlieffen bis zum Beginn des Ersten Weltkrieges," in *Operatives Denken und Handeln in deutschen Streitkräften im 19. and 20. Jahrhundert. Vorträge zur Militärgeschichte*, vol. 9, ed. Militärgeschichtliches Forschungsamt (Herford, 1988), pp. 45–72.

35 Allan Mitchell, *Victors and Vanquished: The German Influence on Army and Church in France after 1870* (Chapel Hill, N.C., 1984) presents this process in detail.

36 The Anglo-French entente of 1904 also generated the "hostage theory," by which, in case of war with England, decisive pressure was to be exerted on an otherwise inaccessible enemy by overrunning France and arguably the Low Countries as well. See Einem to Bülow, 17 October 1904, with Schlieffen's enclosure of 7 October, in PAAA, Deutschland 138 Geheim, vol. 6; and Bülow to Holstein, 15 December 1904, in Friedrich von Holstein, *The Holstein Papers: The Memoirs, Diaries and Correspondence of Friedrich von Holstein, 1837–1909*, ed. N. Rich and M. H. Fisher, vol. 4 (Cambridge, 1963), no. 869.

37 The protocol of the conference was edited by Werner Knoll and Hermann Rahne as "Bedeutung und Aufgaben der Konferenz der Generalstabschefs der Armeekorps in Frankfurt a.M. am 21. Januar 1914," *Militärgeschichte* 25 (1986), pp. 55–63.

38 Wayne C. Thompson, "The September Program: Reflections on the Evidence," *Central European History* 11 (1978), pp. 348–354, and Egmont Zechlin, "Deutschland zwischen Kabinettskrieg und Wirtschaftskrieg: Politik und Kriegführung in den ersten Monaten des Weltkrieges 1914," *Historische Zeitschrift* 199 (1964), pp. 376–381, successfully establish the confused environment in which German war aims were developed.

39 See particularly Adolf Wild von Hohenborn, *Briefe und Tag-
   ebuchaufzeichnungen des Preussischen Generals als Kriegsminister und
   Truppenführer im Ersten Weltkrieg*, ed. H. Reichold and G. Gramer
   (Boppard, 1986). For the consequences of this mind-set, see, inter alia,
   Werner Conze, *Polnische Nation und deutsche Politik im Ersten Weltkrieg*
   (Koln, 1958); Immanuel Geiss, *Der polnische Grenzstreifen 1914–1918*
   (Lübeck, 1960); and Winfried Baumgart, "Das 'Kaspi-Unternehmen.'
   Grössenwahn Ludendorffs oder Routinesplanung des deutschen Gener-
   alstabs," *Jahrbücher für Geschichte Osteuropas* 18 (1970), pp. 47–126,
   231–278.

40 A point well made in Paddy Griffith, *Forward into Battle: Fighting Tactics
   from Waterloo to Vietnam* (New York, 1981), p. 43.

41 Erzberger's comment, part of a speech delivered in the Reichstag on 20
   March 1918, is cited in Wolfram Wette, "Reichstag und 'Kriegsgewinnlerei'
   (1916–1918). Die Anfänge parlamentarischer Rustungskontrolle in
   Deutschland," *Militärgeschichtliche Mitteilungen* 36 (1984), pp. 31–56. See
   also Ursula Ratz, "Sozialdemokratische Arbeiterbewegung, Bürgerliche
   Sozialreformer und Militärbehorden im Ersten Weltkrieg," ibid. 37 (1985),
   pp. 9–33; Gerald D. Feldman, *Army, Industry and Labor in Germany*
   (Princeton, N.J., 1966); K. Koszyk, *Deutsche Pressepolitik im Ersten
   Weltkrieg* (Düsseldorf, 1968); and Jürgen Kocka, *Facing Total War: German
   Society, 1914–1918*, trans. B. Weinberger (Cambridge, Mass., 1984).

42 See the discussion of *Hamsterei* in "Aufzeichnungen des Chefs der
   Fabrikenabteilung des Reichsmarineamts über eine Besprechung im Krieg-
   samt aus Anlass der Streikbewegung," 26 April 1917, in Deist, *Innenpolitik*, vol. 2, pp. 733–734. Karl Ay, *Die Entstehung einer Revolution: Die
   Volksstimmung in Bayern während des Ersten Weltkrieges* (Berlin, 1968)
   incorporates numerous examples of military becoming a focal point for a
   spectrum of domestic grievances. For the radicalization of the army, see
   Dieter Dreetz, "Die Funktionen des Heimatheeres des deutschen Imperialismus während des ersten Weltkrieges und in der Novemberrevolution"
   (Diss., Potsdam, 1975); and, more specifically, Ernst-Heinrich Schmidt,
   *Heimatheer und Revolution 1918. Die militärischen Gewalten im
   Heimatheer zwischen Oktoberreform und Novemberrevolution* (Stuttgart,
   1981).

43 See Martin Kitchen, "Militarism and the Development of Fascist Ideology:
   The Political Ideas of Colonel Max Bauer, 1916–1918," *Central European
   History* 8 (1975), pp. 199–220; and Thoss, "Nationale Rechte."

44 "Schreiben des Oberbefehlshabers der Ostseestreitkräfte an die Immediat-
   behörden der Marine uber den Beitritt von Marineoffizieren zur Deutschen
   Vaterlandspartei," 16 September 1917, and "Erlass des preussischen
   Kriegsministeriums an die Militärbehörden der Heimat betr. die Deutsche
   Vaterlandspartei," 20 November 1917, in Deist, *Innenpolitik*, vol. 2, pp.
   1048ff. and 1101ff.; and entry of 4 November 1918 in *Adjutant im Preussischen Kriegsministerium Juni 1918 bis Oktober 1919. Aufzeichnungen
   des Hauptmanns Gustav Böhm*, ed. H. Hürten and G. Meyer (Stuttgart,
   1977), p. 54.

45 See "Schreiben des preussischen Kriegsministeriums an die Militär-
   befehlshaber betr. Mitgliederwerbung politischer Vereine in Heer und
   Marine," 12 November 1917; and "Weisung des bayerischen
   Kriegsministeriums an den stellv. kommandierenden General des III.
   bayerischen AK betr. die politische Betätigung von Militärpersonen," 27
   December 1917, ibid., pp. 1097ff. and 1117ff.

46 Entry of 12 December 1918 in Böhm, *Adjudant*, p. 93; "Auszüge aus einen
   Aufzeichnungen des Obersten Metz v. Quirnheim . . . ," Deist, *Innen-
   politik*, vol. 2, pp. 783ff.; and the comments of Colonel Haeften of Oberste
   Heeresleitung and former War Minister Karl von Einem, ibid., p. 1344;
   Ludwig Beck to his sister-in-law Gertrud Beck, 28 November 1918, in
   Klaus-Jürgen Müler, *General Ludwig Beck* (Boppard, 1980), pp. 323ff.

47 Wilhelm Meier-Dörnberg, "Die grosse deutsche Frühjahrsoffensive 1918
   zwischen Strategie und Taktik," in *Operatives Denken*, pp. 73–96. Crown
   Prince Rupprecht von Bayern, *Mein Kriegstagebuch*, vol. 2 (Berlin, 1929),
   p. 372; Paul von Hindenburg, *Aus meinem Leben* (Leipzig, 1934), pp. 233ff.

48 See Lupfer, *Dynamics of Doctrine*, pp. 38ff.; Gudmundsson, *Forlorn Hope*,
   passim; Bryan Perrett, *A History of Blitzkrieg* (New York, 1983), pp. 28ff.;
   and Laszlo Alföldi, "The Hutier Legend," *Parameters* 5 (1976), pp. 69–74.

49 For a brief analysis of the German army's exhaustion and decline, see
   Wilhelm Deist, "Der militärische Zusammenbruch des Kaiserreichs," in
   *Das Unrechtsregime. Internationale Forschungen uber den National-
   sozialismus*, ed. U. Buttner, vol. 1 (Hamburg, 1986), pp. 101–129.

50 Wolfram Wette, *Gustav Noske. Eine politische Biographie* (Düsseldorf,
   1987), pp. 263ff.

51 On the issue of modernization, see, most recently, Gerald D. Feldman, "The
   Weimar Republic: A Problem of Modernization?" *Archiv für
   Sozialgeschichte* 26 (1986), pp. 1–26; and Knut Borchardt, *Wachstum,
   Krisen, Handlungsspielräume der Wirtschaftspolitik. Studien zur
   Wirtschaftsgeschichte des 19. und 20. Jahrhunderts* (Göttingen, 1982), pp.
   165–205. For the military aspects, see Ernst Hansen, *Reichswehr und
   Industrie* (Boppard, 1978); Gaines Post, Jr., *The Civil-Military Fabric of
   Weimar Foreign Policy* (Princeton, N.J., 1973); and Rolf-Dieter Müller, *Das
   Tor zur Weltmacht. Die Bedeutung der Sowjetunion für die deutsche
   Wirtschafts-und Rüstungspolitik zwischen den Weltkriegen* (Boppard,
   1984).

52 Robert M. Citino, *The Evolution of Blitzkrieg Tatics: Germany Defends
   Itself against Poland, 1918–1933* (Westport, Conn., 1987) is a mistitled but
   useful case study of the military elements of Weimar Germany's security
   problem in one crucial area.

53 Two recent surveys of this subject, incorporating comprehensive discussions of the literature, are P. Krüger, *Die Aussenpolitik der Republik von
   Weimar* (Darmstadt, 1985) and Marshall M. Lee and Wolfgang Michalka,
   German Foreign Policy, 1917–1933: Continuity or Break? (New York,
   1987). Ulrich Heinemann, *Die verdrängte Niderlage. Politische Öf-
   fentlichkeit und Kriegsschuldfrage in der Weimarer Republik* (Göttingen,
   1983) stresses the extension of revisionist sentiment deep into the German
   left.

54 These alternatives are developed in Michael Geyer, *Aufrüstung oder
   Sicherheit. Reichswehr in der Krise der Machtpolitik* (Wiesbaden, 1980);
   and "The Dynamics of Military Revisionism in the Interwar Years: Military
   Politics between Rearmament and Diplomacy," in *The German Military in
   the Age of Total War*, ed. W. Deist (Dover, N.H., 1988), pp. 100–151.

55 See Klaus-Jürgen Müller, "The Army and the Third Reich: An Essay in
   Historical Interpretation," in *The Army, Politics and Society in Germany,
   1933–45* (Manchester, 1987), pp. 16–53; and *Armee und Drittes Reich
   1933–1939*.Older but still useful are Robert J. O'Neill, *The German Army
   and the Nazi Party, 1936–1939* (London, 1966); and Manfred Mes-

serschmidt, *Die Wehrmacht im NS-Staat. Zeit der Indoktrination* (Hamburg, 1969).

56 See the detailed discussion of this issue in Dennis E. Showalter, "The Political Soldiers of Imperial Germany: Myths and Realities," in *Politics, Parties, and the Authoritarian State: Imperial Germany, 1871–1918*, vol. 1, ed. John Fout (forthcoming).

57 For aspects of this problem, see Hans Meier-Welcker, *Seeckt* (Frankfurt, 1967), pp. 218ff.; Heinz Sperling, "Die Tätigkeit und Wirksamkeit des Heereswaffenamtes der Reichswehr für die materiell-technische Ausstattung eines 21-Divisionen Heeres 1924–1934" (Diss., Potsdam, 1980); and David N. Spires, *Image and Reality: The Making of the German Officer, 1921–1933* (Westport, Conn., 1984).

58 See inter alia, Paul Koistinen, *The Hammer and the Sword: Labor, the Military, and Industrial Mobilization, 1920–1945* (New York, 1979); "The 'Industrial-Military Complex' in Historical Perspective: The Interwar Years," *The Journal of American History* 56 (1970), pp. 819–839; and D. K. R. Crosswell, "'Aides, Adjutants, and Asses'—The United States Army's Advanced Schools in the Inter-War Years," presented at the 1989 meeting of the American Military Institute. See also such general accounts as Russell F. Weigley, *The American Way of War: A History of United States Military Strategy and Policy* (New York, 1973), pp. 203ff.; and I. B. Holley, *General John M. Palmer: Citizen Soldiers and the Army of a Democracy* (Westport, Conn., 1982), pp. 413ff.

59 See particularly two significant *Dokumentationen:* Günter Wollstein, "Eine Denkschrift des Staatssekretärs Bernhard von Bülow von März 1933. Wilhelminische Konzeption der Aussenpolitik zu Beginn der nationalsozialistischen Herrschaft," *Militärgeschichtliche Mitteilungen* 13 (1973), pp. 77–94; and Hans-Jürgen Rautenberg, "Drei Dokumente zur Planung eines 300,000-Mann-Friedensheeres aus dem Dezember 1933," ibid. 22 (1977), pp. 103–139.

60 "Aufzeichnungen Holtzmanns über einen Besuch bei Beck am 16.11.1938" in Müller, *Beck*, p. 579. Since Holtzmann was a confidant of Ludendorff's, Beck might be suspected of flattery, but see ibid., p. 80, and Geyer, "Military Revisionism," pp. 133ff.

61 See Wilhelm Deist, *The Wehrmacht and German Rearmament* (London, 1981); Fritz Blaich, "Wirtschaft und Rüstung in Deutschland 1933–1939," in *Sommer 1939. Die Grossmächte und der europäische Krieg*, ed. W. Benz and H. Graml (Stuttgart, 1979), pp. 35–59; Michael Geyer, "Rüstungsbeschleunigung und Inflation. Zur Inflationsdenkschrift des Oberkommandes der Wehrmacht von November 1938," *Militärgeschichtliche Mitteilungen* 30 (1981), pp. 121–186; Ludolf Herbst, "Die Krise des Nationalsozialistischen Regimes am Vorabend des Zweiten Weltkrieges und die forcierte Aufrüstung. Eine Kritik," *Vierteljahreshefte für Zeitgeschichte* 26 (1978), pp. 347 392; and Richard Overy, *Goering: The 'Iron Man'* (London, 1984), pp. 48–108. The latter work, misleadingly titled apparently for market purposes, is a first-rate analysis of the Nazi economy at war.

62 A point well developed in K. Krafft von Delmensingen, *Der Durchbruch. Studie an Hand der Vorgänge des Weltkrieges 1914–1918* (Hamburg, 1937), pp. 132ff.

63 See, inter alia, Beck's memorandum of 30 December 1935 on the subject of improving the army's offensive power, in Müller, *Beck*, pp. 469ff.; Heinz

Guderian, *Panzer Leader* (New York, 1957), pp. 20ff.; and Williamson K. Murray, "German Army Doctrine, 1918–1939, and the Post-1945 Theory of 'Blitzkrieg Strategy,'" in *German Nationalism and the European Response, 1890–1945*, ed. C. Fink et al. (Norman, Okla., 1985), pp. 71–94.

64 The best overview is Alf Lüdtke, "'Wehrhafte Nation' und 'innere Wohlfart.' Zur militärischen Mobilisierbarkeit der bürgerlichen Gesellschaft, Konflikt und Konsens zwischen Militär und ziviler Administration in Preussen," *Militärgeschichtliche Mitteilungen* 30 (1981), pp. 7–56.

65 Manfred Messerschmidt, "The Wehrmacht and the Volksgemeinschaft," *Journal of Contemporary History* 18 (1983), pp. 719–749; and A. Schildt, *Militärdiktatur mit Massenbasis? Die Querfrontkonzeption der Reichswehrführung um General Schleicher am Ende der Weimarer Republik* (Frankfurt, 1981).

66 See Geoffrey Stoakes, *Hitler and the Quest for World Dominion* (New York, 1986); Klaus Hildebrand, *Deutsche Aussenpolitik 1933–1945. Kalkül oder Dogma*, 4th ed. with an afterword, "Die Geschichte der deutschen Aussenpolitik (1933–1945) im Urteil der neueren Forschung: Ergebnisse, Kontroversen, Perspectiven" (Stuttgart, 1980); Manfred Messerschmidt, "Aussenpolitik und Kriegsvorbereitung," in *Das Deutsche Reich und der Zweite Weltkrieg*, vol. 1, *Ursachen und Voraussetzungen der deutschen Kriegspolitik* (Stuttgart, 1979), pp. 535–701; and, for the impact of Munich and its aftermath on the army, Williamson K. Murray, *The Change in the European Balance of Power, 1938–1939: The Path to Ruin* (Princeton, N.J., 1984).

67 See Harold C. Deutsch, *The Conspiracy against Hitler in the Twilight War* (Minneapolis, 1968); and Klaus-Jürgen Müller, "Struktur und Entwicklung der nationalkonservativen Opposition in Deutschland," in *Aufstand des Gewissens. Militärischer Widerstand gegen Hitler und das NS-Regime 1933–1945*, rev. ed., Militärgeschichtliches-Forschungsamt (Herford and Bonn, 1985), pp. 263–309.

68 On the early relationship of the army to the Waffen-SS, see particularly Bernd Wegner, *Hitlers Politische Soldaten: Die Waffen-SS 1933–1945*, 2d ed. (Paderborn, 1983). Older but still useful are George H. Stein, *The Waffen-SS: Hitler's Elite Guard at War* (Ithaca, N.Y., 1966) and James J. Weingartner, *Hitler's Guard: The Story of the Leibstandante SS Adolf Hitler, 1933–1945* (Carbondale, Ill., 1974), pp. 11ff.

69 See Beck's observations on the Hossbach memorandum of 12 November 1937; and his "Observations on the Military-Political Situation" of 5 May 1938, in Müller, *Beck*, pp. 498ff. and 502ff. See also the diary entries of 14, 22, 28, and 31 August 1939, in *The Halder War Diary, 1939–1942*, ed. C. Burdick and H.-A. Jacobsen (Novato, Calif., 1988), pp. 11ff., 29ff., 37ff., and 43–44.

70 Entries of 17 and 30 March 1941, Burdick and Jacobsen, *Halder War Diary*, pp. 833, 845–846. See Timotny Patrick Mulligan, *The Politics of Illusion and Empire: German Occupation Policy in the Soviet Union, 1942–1943* (New York, 1988); and Theo Schulte, *The German Army and Nazi Policies in Occupied Russia, 1941–1945* (Oxford, 1989).

71 See Peter Hoffmann, "Der militärische Widerstand in der zweiten Kriegshälfte," in *Aufstand des Gewissens*, pp. 395–419; and Manfred Messerschmidt, "Militärische Motive zur Durchführung des Umsturzes," in *Der Widerstand gegen den Nationalsozialismus*, ed. J. Schmadeke and P. Steinbach (Munich, 1985), pp. 1021–1036.

72 This issue is explored in, inter alia, Christian Streit, *Keine Kameraden. Die Wehrmacht und die Sowjetischen Kriegsgefangenen 1941–1945* (Stuttgart, 1978); Omer Bartov, *The Eastern Front, 1941–45: German Troops and the Barbarization of Warfare* (New York, 1986); and Jürgen Forster, "New Wine in Old Skins? The Wehrmacht and the War of 'Weltanschauungen,' 1941," in *German Military in the Age of Total War*, pp. 304–322.

73 On this subject, see Eberhard Schwarz, *Die Stabilisierung der Ostfront nach Stalingrad. Mansteins Gegenschlag zwischen Donez und Dnjepr im Fruhjahr 1943* (Göttingen, 1986).

Chapter Eight. Arms and Alliances

1 J. C. Cairns, "Some Recent Historians and the 'Strange Defeat' of 1940", *Journal of Modern History* no. 1 (March 1974), p. 75.

2 Marc Bloch, *Strange Defeat* (London, 1949), pp. 125, 135, 138, 140, 156, 168.

3 See, for example, R. D. Anderson, *France, 1870–1914: Politics and Society* (London, 1977).

4 Jean-Pierre Azéma, *From Munich to the Liberation, 1938–1944* (Cambridge, 1984), pp. 15–17.

5 Robert Frankenstein, *Le Prix de réarmament français, 1935–1939* (Paris, 1982), p. 289.

6 Robert Doughty, *The Seeds of Disaster: The Development of French Army Doctrine, 1919–1939* (Hamden, Conn., 1985), p. 183.

7 Ibid., pp. 179–181.

8 Brian Bond and Martin Alexander, "Liddell Hart and De Gaulle: The Doctrines of Limited Liability and Mobile Defense", in Peter Paret, ed., *Makers of Modern Strategy* (Princeton, N.J., 1986), pp. 610, 615.

9 Gerd Krumeich, *Armament and Politics in France on the Eve of the First World War: The Introduction of Three-Year Conscription* (Dover, N.H., 1985), p. 18.

10 Michael Howard, 'Men against Fire: The Doctrine of the Offensive of 1914", in Paret, *Makers of Modern Strategy*, pp. 522–523.

11 Service historique de l'armée de terre, château de Vincennes, 7N 1538, December 1911 note of Colonel Janin of 2ᵉ bureau.

12 Joseph Joffre, *Memoires du Maréchal Joffre, 1910–1917*, vol. 1 (Paris, 1932), pp. 131–134. Krumeich concedes that the actual implementation of formal military agreements between the French and Russian General Staffs was a constant source of difficulty, but he rejects as "hardly appropriate" S. R. Williamson's view that this was due to a "latent mistrust" between the two groups (Krumeich, *Armaments*, p. 271, note 11). But this mistrust did exist, and clouded the French appreciation of Russian intentions in the event of war. Even Krumeich admits that the Russians were only convinced that war with Germany was virtually imminent by the Liman von Sanders crisis of January 1914 (ibid., p. 124). Therefore, at the very least, Plan XVII was drawn up by the French in the absence of hard evidence of Russian intentions.

13 S. R. Williamson, *The Politics of the Grand Strategy* (Cambridge, Mass., 1969), p. 226.

14 William C. Fuller, Jr., "The Russian Empire," in Ernest R. May, ed.,

Knowing One's Enemies: Intelligence Assessment before the Two World Wars (Princeton, N.J., 1984), pp. 100–104, 110–112, 122–125.

15 Williamson Murray, *The Change in the European Balance of Power, 1938–1939: The Path to Ruin* (Princeton, N.J., 1984), p. 309.

16 Jean-Jacques Becker, *1914: Comment les Français sont entrés dans la guerre* (Paris, 1977), pp. 557–558, 575, 582, 587.

17 Douglas Porch, "Bugeaud, Galliéni, Lyautey. The Development of French Colonial Warfare", in Paret, *Makers of Modern Strategy*, p. 405.

18 Joffre, *Memoires*, vol. 1, p. 104.

19 Christopher M. Andrew, "France and the German Menace", in May, *Knowing One's Enemies*, pp. 139, 143.

20 Williamson, *Politics*, pp. 222–223.

21 Andrew, "France and the German Menace," pp. 148–149.

22 Williamson, *Politics*, pp. 218.

23 Krumeich, *Armaments*, p. 27.

24 Joffre, *Memoires*, vol. 1, p. 130.

25 Jiri Hochman, *The Soviet Union and the Failure of Collective Security, 1934–1938* (Ithaca, N.Y., 1984), pp. 54–55, 76–77.

26 Ibid.

27 Doughty, *Seeds of Disaster*, pp. 181–182.

28 Murray, *Change*, pp. 311–314, 328, 341.

29 Jack Snyder, *The Ideology of the Offensive: Military Decision Making and the Disasters of 1914* (Ithaca, N.Y., 1984), pp. 10, 18, 24–25, 27, 97.

30 Barry Posen, *The Sources of Military Doctrine: France, Britain, and Germany between the World Wars* (Ithaca, N.Y., 1984), pp. 118–119.

31 Douglas Porch, *The March to the Marne: The French Army, 1871–1914* (Cambridge, 1981), chap. 11.

32 Philip Bankwitz, *Maxime Weygand and Civil-Military Relations in Modern France* (Cambridge, Mass., 1967), pp. 45–46.

33 Porch, *March to the Marne*, pp. 243–244. Krumeich, *Armaments*, pp. 45, 49

34 Porch, *March to the Marne*.

35 Richard D. Challener, *The French Theory of the Nation in Arms, 1866–1931* (New York, 1955), p. 86.

36 André Beaufre, *1940: The Fall of France* (New York, 1968), p. 157. See Doughty, *Seeds of Disaster*, chap. 9, for the best recent discussion of the problems of formulating a doctrine in the inter-war French army. Challener, *French Theory*, chap. 9, examines the influence of recruitment policy on the lack of military innovation.

Chapter Nine. The Evolution of Soviet Grand Strategy

1 Some idea of the Marxist conception of war can be gained by perusing two English-language collections: Bernard Semmel, ed., *Marxism and the Science of War* (Oxford, 1981), and *Marxism-Leninism on War and Army* (Moscow, 1972), which was written by a "collective" of Soviet military authors.

2 Marshal V. D. Sokolovskii, "Some Problems of Military Science," *Military Thought*, September 1964.

3 Alexander Dallin has called this a conflict between "the desire to enjoy and the need to destroy."

4 One explanation for how a revolutionary government created a nonrevolutionary army, is given by Leon Trotsky in *My Life: An Attempt at an Autobiography* (New York, 1970), p. 438, and idem, "Military Doctrine or Pseudo-Military Doctrinairism," in *Military Writings* (New York, 1971), pp. 34–36.

5 M. N. Tukhachevsky, "Letter to G. Zinoviev, 18 July 1920," repr. in John Erickson, *The Soviet High Command: A Military-Political History, 1918–1941* (New York, 1962), pp. 784–785.

6 See, for instance, Trotsky's polemic against exporting revolution, in his "Military Doctrine or Pseudo-Military Doctrinairism."

6 Stalin's first public espousal of "socialism in one country" was through an article entitled "The October Revolution and the Tactics of the Russian Communists," which appeared as the preface to his book *On the Road to October*.

8 I. V. Stalin, *On the Opposition* (Peking, 1974), p. 325.

9 The need for "socialism in one country" tied to a strong military is discussed most clearly by Stalin in "Ob oppozitsionnom bloke v VKP(b)" and "O sotsial-demokraticheskom uklone v nashei partii," both in I. V. Stalin, *Sochineniia. Tom 8, 1926: Yanvar'-Noyabr'.* (Moscow, 1948).

10 The influence of the "rightists," including Tomsky, Rykov, and Bukharin, is examined thoroughly in Stephen F. Cohen, *Bukharin and the Bolshevik Revolution: A Political Biography, 1888–1938* (Oxford, 1980).

11 Robert Conquest discusses this very well in *The Great Terror: Stalin's Purge of the Thirties* (New York, 1973).

12 M. N. Tukhachevsky, *Izbrannyye Proizvedeniya. Tom Pervyy: 1919–1927* (Moscow, 1964), p. 12.

13 The officers most involved in creating the concept of deep operations were Tukhachevisky, A. I. Yegorov, and V. K. Triandafillov. See N. V. Ogarkov, "Glubokaya Operatsiya," in *Sovetskaya Voyennaya Entsiklopediya* (Moscow, 1976–1980), pp. 574–578.

14 For the complete story, see pp. 449–509 of Erickson, *Soviet High Command.*

15 One of the earliest discussions of Stalin's wartime mistakes can be found in a Party history, Ministerstva Oborony Soyuza SSR, *Istoriya velikoi otechestvennoi voiny sovetskovo soyuza 1941–1945* (Moscow, 1960). Glasnost has permitted an even more open forum on the role Stalin played as a military leader during World War II. Especially interesting are Dmitry Volkogonov's biography of Stalin, *Triumf i tragedia* (Moscow, 1989), and such articles as G. Kumanev's "Razmyshleniya istoika: 22-vo, na rassvete," *Pravda*, 22 June 1989, p. 3.

16 John Lewis Gaddis, *The United States and the Origins of the Cold War, 1941–1947* (New York, 1972), pp. 353–361. Gaddis includes an excellent overview of the debate concerning the origins of the Cold War.

17 N. S. Khrushchev, *Report of the Central Committee to the 20th Congress of the CPSU* (London, 1956), p. 28. That this doctrine was continued long after Khrushchev was gone can be seen in V. D. Sokolovskii's declaration

that "Marxism-Leninism teaches that socialist revolutions do not necessarily involve war. . . . The great aims of the working class in the present
era can be accomplished without world war and without civil war—by
peaceful means." V. D. Sokolovskii et al., *Soviet Military Strategy* (New
York, 1975), p. 182.

18 Uri Ra'anan, *The USSR Arms the Third World: Case Studies in Soviet
Foreign Policy* (Cambridge, Mass., 1969), pp. 13–158.

19 Among other statements of this policy, see Khrushchev's "Concluding
Speech Delivered at the Twenty-First Extraordinary Congress of the CPSU,"
in N. S. Khrushchov [sic], *World without Arms, World without Wars,* Book
1: January–July 1959 (Moscow, n.d.), pp. 65–66.

20 For the impact of the Sino-Soviet conflict on Soviet third world policies,
see "Chinese Party's '25 Points' (June 1963)"; "Soviet Reply to the '25
Points' (July 1963)"; and "Mikhail Suslov's Report to the CPSU Central
Committee (February 1964)," all in *China and the Soviet Union*, compiled
by Peter Jones and Sian Kevill and edited by Alan J. Day (New York, 1985),
pp. 31–44 and 49–50.

21 Massive retaliation was first enunciated in 1953 through NSC 162-2, the
policy statement which revised Truman's NSC 68. See Norman A.
Graebner, ed., *The National Security: Its Theory and Practice, 1945–1960*
(Oxford, 1986), p. 51.

22 Robert P. Berman and John C. Baker, *Soviet Strategic Forces: Requirements
and Responses* (Washington, D.C., 1982), pp. 46–47, 55.

23 The viewpoint that nuclear weapons had permanently altered warfare, and
the opposing view—that nuclear weapons were simply more powerful
artillery pieces—can both be found in Sokolovskii et al., *Soviet Military
Strategy*, pp. 170–171, 191–194, and 241–242.

24 David Holloway, *The Soviet Union and the Arms Race* (New Haven, 1983),
pp. 37–41.

25 V. F. Tolubko, *Nedelin* (Moscow, 1979), pp. 187–188.

26 M. V. Zakharov, "Vlastonoye trebovaniye vremeni," *Krasnaya zvezda*, 4
February 1965, pp. 2–3.

27 Coit D. Blacker. "The Kremlin and Détente: Soviet Conceptions, Hopes,
and Expectations," in Alexander L. George, ed. *Managing U.S.-Soviet
Rivalry: Problems of Crisis Prevention* (Boulder, Colo., 1983), pp. 120–121.

28 L. I. Brezhnev, "Report of the CPSU Central Committee to the Twenty-
Fourth Congress of the Communist Party of the Soviet Union," in *Following
Lenin's Course: Speeches and Articles* (Moscow, 1972), p. 355.

29 Mikhail Gorbachev, "Politicheskii doklad tsentral'novo komiteta KPSS
XXVII c'yezdu kommunisticheskoi partii Sovetskogo Soyuza," *Pravda*, 26
February 1986.

30 This is being publicly recognized in the Soviet Union as well. See Ye.
Primakov, "Novaya filosofiya vneshnei politiki," *Pravda*, 7 July 1988, p. 4.
In this article, Primakov comes as close as any top Soviet leader has to
making a direct linkage between the need to reduce military spending and
the worsening economy.

31 One of the most interesting discussions of the shortcomings of Brezhnev's
foreign policy comes from a roundtable sponsored by and reported in
Literaturnaya Gazeta, 29 June 1988.

32 See Condoleeza Rice, "The Party, the Military, and Decision Authority in the Soviet Union," *World Politics,* October 1987, pp. 55–81.

33 See, for example, M. A. Gareyev, *M. V. Frunze: Voennyi teoretik* (Moscow, 1985), pp. 242–243. This is also discussed in Abraham Becker, *Ogarkov's Complaint and Gorbachev's Dilemma: The Soviet Defense Budget and Party-Military Conflict* (Santa Monica, Calif., 1987).

34 The debate over "defense conversion" is presented well in an article by A. Kireyev, "Conversion to Economic Accountability," *Ogonëk* 27 (July 1989), pp. 6–7, 26–27 (Foreign Broadcast Information Service translation).

35 "Soviets Seek GATT Entry," *The New York Times,* 14 March 1989, p. D26, and S. Karen Witcher, "Soviets Consider Joining IMF, World Bank," *The Wall Street Journal,* 15 August 1986, p. 19.

36 TASS, "Za delo-bez raskachki," *Pravda,* 6 August 1988, p. 2, and Bill Keller, "Gorbachev Deputy Criticizes Policy," *The New York Times,* 7 August 1988, sec. 1, p. 11.

Chapter Ten. American Grand Strategy, Today and Tomorrow

1 See chap. 1 above.

2 Apart from John Elliott's essay in this collection, see Geoffrey Parker, *Spain and the Netherlands, 1559–1659* (London, 1979).

3 Apart from the essays by Michael Howard and Eliot Cohen in this volume, see Paul Kennedy, *Strategy and Diplomacy, 1870–1945: Eight Studies* (London, 1983), chaps. 2 and 3.

4 See chap. 1 above.

5 I am grateful to Eliot Cohen for insisting that I emphasize this point, and not "blur over" the very different conditions between war and peace in my musings about the essential nature of grand strategy. One of the intellectual—and methodological—difficulties is that much of the historical literature has focused *either* upon wartime *or* upon peacetime grand strategy, rather than analyzing the continuities and discontinues between the two.

6 See chap. 1 above.

7 This attitude, that the West had to avoid the mistakes of the "appeasers," is brought out very clearly in Ernest R. May, *"Lessons" of the Past: The Use and Misuse of History in American Foreign Policy* (New York, 1973), pp. 50–51; and in the angry reaction to the 1961 publication of A. J. P. Taylor's *The Origins of the Second World War,* which was seen by some critics as being not only an apologia for Hitler, but also for Khrushchev's current policies; see C. R. Cole, "Critics of the Taylor View of History," in Esmonde M. Robertson, ed., *The Origins of the Second World War: Historical Interpretations* (London, 1971), p. 155.

8 This debate is covered in, inter alia, Geoffrey R. Searle, *The Quest for National Efficiency, 1899–1914* (Oxford, 1971), and Paul Kennedy, *The Rise of the Anglo-German Antagonism, 1860–1914* (London and Boston, 1980), chaps. 16–18.

9 For a sampling, see James Fallows, *National Defense* (New York, 1981); Robert W. DeGrasse, *Military Expansion, Economic Decline* (Armonk, N.Y., 1985); Mary Kaldor, *The Baroque Arsenal* (London, 1983); and Richard Rosecrance, *The Rise of the Trading State* (New York, 1985).

10 Even *without* those conflicts, the European-centered world was being steadily eclipsed by the rise of the United States and Russia; but the world wars certainly helped to accelerate that trend; see Geoffrey Barraclough, *An Introduction to Contemporary History* (Harmondsworth, 1967), chaps. 3 and 4; Paul Kennedy, *The Rise and Fall of the Great Powers* (New York, 1987), chaps. 5 and 6.

11 Ronald Steel, *Pax Americana*, rev. ed. (Harmondsworth, 1970), chap. 2; Robert Dallek, "The Postwar World: Made in the USA," in Sanford J. Ungar, *Estrangement: America and the World* (New York, 1985), pp. 29–49.

12 For some statistical evidence, see Kennedy, *Rise and Fall*, pp. 414–415.

13 Quoted in Kenneth Bourne, *The Foreign Policy of Victorian England, 1830–1902* (Oxford, 1970), p. 409.

14 One thinks here not only of Roosevelt's confident call in 1940 for an American production capacity of "fifth thousand planes a year" (see Michael Sherry, *The Rise of American Air Power* (New Haven, 1987), p. 91); and of Churchill's conviction that the United States economy was like a great machine, or a "gigantic boiler," with no limits to its productive output (see Eliot Cohen's essay in this collection, and David Reynolds, *The Creation of the Anglo-American Alliance 1937–41* [London, 1981]).

15 See John Gaddis, *Strategies of Containment* (New York, 1982), p. 59.

16 This refers to James Joll's penetrating essay "1914: The Unspoken Assumptions," which is reproduced in Hansjoachim W. Koch, ed., *The Origins of the First World War* (London, 1972), chap. 8.

17 This is the entire first paragraph of the commission's report, entitled *Discriminate Deterrence* (Washington, D.C., 1988), p. 5.

18 There are useful surveys by Samuel R. Huntington, "The U.S.—Decline or Renewal?" *Foreign Affairs* 67, no. 1 (Winter 1988–1989), pp. 76–96; and D. Reynolds, "Power, Wealth and War in the Modern World," *Historical Journal* 33, no. 2 (June 1989), pp. 475–487.

19 For some further details of these developments, see Kennedy, *Rise and Fall*, chap. 8.

20 See the graph in *Discriminate Deterrence*, p. 7 (which does not include the European Community).

21 Here, clearly, it is the Japanese-American relationship which has altered the most over that period. For example, although Japan still relies upon American military protection, it is the United States which nowadays relies upon Japanese capital flows, and upon a great deal of Japanese technology—as Japanese conservatives proudly point out; see the text of Akio Morita and Shintaro Ishihara, *The Japan That Can Say "No": The New U.S.-Japan Relations Card ["No" to Ieru Nippon: Shin Nichi-Bei Kankei no Kaado]* (Tokyo, 1989).

22 It may be worth making the obvious remark here that certain other aspects have not changed very much or at all (and therefore attract far less attention): for example, the commitment of both Republican and Democratic parties to the alliance system, the ongoing public debate between the "realist" and "idealist" views of foreign policy, the problems of interservice cooperation, and the efforts to reform defense procurement.

23 How, for example, could the present size of many of the weapons systems of the U.S. Navy be justified if the Soviet fleet were greatly reduced in numbers and operations? As this paragraph was being drafted, news

arrived of the USSR's intention to scrap large numbers of its surface warships and submarines (*The Wall Street Journal*, 21 July 1989, p. A12), and of its plan to reduce tank production enormously (*The New York Times*, 22 July 1989, p. 6). The latter article is placed (ironically or deliberately?) next to a report on the U.S. Air Force's pleas to save funding for the B-2 ("Stealth") bomber.

24 See the analyses by L. Sullivan, Jr., "Major Defense Budget Costs," in *Defense Economics for the 1990s: Resources, Strategies, and Options* (Washington, D.C., 1989).

25 See the coverage in Jean-Baptiste Duroselle, *La Décadence, 1933–1939* (Paris, 1979); and in Kennedy, *The Realities behind Diplomacy* (London, 1981), chaps. 5 and 6.

26 As represented in the *Discriminate Deterrence* report, from which (p. 1) the quotations are taken.

27 The quotation is from Brodie's *The Absolute Weapon*. See Lawrence Freedman, *The Evolution of Nuclear Strategy* (London, 1981), p. 44.

28 Robert Jervis, *The Illogic of American Nuclear Strategy* (Ithaca, N.Y., 1984), p. 19.

29 See Paul Bracken, *The Command and Control of Nuclear Weapons* (New Haven, 1983).

30 David C. Hendrickson, *The Future of American Strategy* (New York, 1989), p. 135. Professor Hendrickson outlines the systems—manned bombers, Trident I submarines, and cruise missiles—which fit that description.

31 See the very good analysis in Brian Bond, *British Military Policy between Two World Wars* (Oxford, 1980).

32 This is well covered in Correlli Barnett, *The Collapse of British Power* (London and New York, 1972), chap. 4.

33 To give but one example: neither Olivares nor his monarch, Philip IV, contemplated sharing "influence" with those parts of the Spanish-Habsburg empire that they were pressing to assume greater financial "burdens" (see John Elliott's essay in this collection).

34 This, I had believed, was the simple point I was making in the (now notorious) section upon the United States in the final chapter of *The Rise and Fall of the Great Powers*, pp. 514–535. It is, essentially, a very conservative viewpoint, because it is about the long-term preservation of the nation's *power* (which may be, as noted above, an anachronistic way of thinking).

35 The phrase refers again to the early-to-mid-1930s British reassessment of their defense "deficiencies" and defense "requirements" in the light of the worsening global situation; see Barnett, *Collapse of British Power*, pp. 342ff.

Acknowledgments

The above essays represent the revised version of a series of public lectures delivered each Monday afternoon in the fall 1987 semester at Yale University. Every year these lectures, which are now funded through the generous assistance of the Lynde and Harry Bradley Foundation, deal with a particular theme in military and strategic history. On this occasion, the speakers were asked to focus upon the largest perspective of all, that is, "grand strategy."

This collection could not have come into being without the aid of a number of people, in particular my colleague Dr. George Andreopoulos, who assisted in innumerable ways. Mary Habeck and Marko Prelec gave invaluable help in the preparation of the finished manuscript, as did Laura Dooley, Harry Haskell, and the staff at Yale University Press. It was a pleasure to work with Charles Grench, executive editor of the Press, and a relief to have the traditional aid and advice of my literary agents, Bruce Hunter and Claire Smith.

Above all, my thanks go to the contributors to this volume, who must have wondered whether their essays would ever see the light of day. I hope they are not dissatisfied with the result.

Paul Kennedy
New Haven, October 1990

Note on Contributors

ELIOT A. COHEN is professor of strategic studies at the Paul H. Nitze School of Advanced International Studies, Johns Hopkins University. His most recent book, written with John Gooch, is *Military Misfortunes: The Anatomy of Failure in War*. He has written widely about contemporary strategic issues, including intelligence, net assessment, and defense organization, and is currently preparing a book on civil-military relations in wartime. He has worked on the Policy Planning Staff of the Office of the Secretary of Defense, and is a consultant to a number of government agencies.

JOHN H. ELLIOTT is the Regius Professor of Modern History in the University of Oxford. Educated at Eton College, he read history at Cambridge, where he subsequently became a Fellow of Trinity College and a University Lecturer in history. In 1968 he was appointed professor of history at King's College, University of London, where he remained until 1973, when he became a professor in the School of Historical Studies at the Institute for Advanced Study in Princeton, N. J. He took up his appointment at Oxford, where he is a Fellow of Oriel College, in July 1990. He has specialized in the history of Spain, Spanish America, and Europe in the sixteenth and seventeenth centuries. Among his publications are *Imperial Spain, 1469–1716*, *The Revolt of the Catalans*, *The Old World and the New, 1492–1650*, (with

Jonathan Brown) *A Palace for a King: The Buen Retiro and the Court of Philip IV, The Count-Duke of Olivares,* and *Spain and Its World, 1500–1700.*

ARTHER FERRILL is professor of history at the University of Washington in Seattle. He is the author of *The Origins of War* and *The Fall of the Roman Empire: The Military Explanation.* He also serves as contributing editor to *MHQ: The Quarterly Journal of Military History* and is the general editor of a series of books published by Westview Press entitled *History and Warfare.* His new books on the Roman emperor Caligula and on Roman imperial grand strategy will be published in 1991. In 1989 he was Solomon Katz Distinguished Lecturer in the Humanities at the University of Washington.

JOHN B. HATTENDORF is the Ernest J. King Professor of Maritime History and Director of the Advanced Research Department at the Naval War College, Newport, R. I. He studied history at Kenyon College, Brown University, and the University of Oxford, served as an officer in the U.S. Navy, and has been visiting professor in military and naval history at the National University of Singapore and at the German Armed Forces Military History Research Office. Most recently, he has coedited *Maritime Strategy and the Balance of Power* and *The Limitations of Military Power.*

SIR MICHAEL HOWARD is Robert A. Lovett Professor of Military and Naval History at Yale University, having formerly held the Regius Chair of Modern History at Oxford. His works include *Grand Strategy,* in the U.K. Official History of the Second World War, and *The Continental Commitment: The Dilemma of British Strategy in the Era of Two World Wars.* His latest publication is *Strategic Deception in the Second World War.*

PAUL KENNEDY is the Dilworth Professor of History at Yale University and director of its International Security Programs. Educated at the universities of Newcastle, Oxford, and Bonn, he taught at the University of East Anglia between 1970 and 1983 before moving to Yale. He was formerly research assistant to Sir Basil Liddell Hart. Among the books that Professor Kennedy has written or edited are *The Rise and Fall of British Naval Mastery, Strategy and Diplomacy, The Rise of the Anglo-German Antagonism, The Realities behind Diplomacy,* and, most recently, *The Rise and Fall of the Great Powers.*

DOUGLAS PORCH received his bachelor's degree from the University of the South, Sewanee, Tenn., in 1967 and his doctorate from Corpus Christi College, Cambridge, in 1972. He is the author of several works on the French army, including *Army and Revolution: France, 1815–1848, The March to the Marne: The French Army, 1871–1914, The Conquest of Morocco, The Conquest of the Sahara,* and *A History of the French Foreign Legion.* He is the Mark W. Clark Professor of History at the Citadel.

CONDOLEEZZA RICE is associate professor of political science at Stanford University. Dr. Rice received her bachelor's degree from the University of Denver, Phi Beta Kappa and Cum Laude; her master's from the University of Notre Dame; and her doctorate from the Graduate School of International Studies at the University of Denver. She is the author of *The Soviet Union and the Czechoslovak Army* and, with Alexander Dallin, *The Gorbachev Era,* as well as numerous articles on Soviet and East European military policy. Dr. Rice has been a Hoover Institution National Fellow and a Council on Foreign Relations Fellow, as well as serving as a consultant to ABC News on Soviet affairs. She has recently returned to Stanford after serving consecutively as the senior director for Soviet affairs on the staff of the National Security Council and then as a special assistant to the President for national security affairs.

DENNIS SHOWALTER is professor of history at Colorado College, Colorado Springs. He is a specialist in German military history. His publications in the field include *Railroads and Rifles: Soldiers, Technology and the Unification of Germany; German Military History, 1648–1982: A Critical Bibliography;* and *Tannenberg: Clash of Empires,* as well as numerous articles and essays on the German way of war.

Index